The Layman's Guide to Making Sense of Statistics

Another Vantage Press title by the author:

Introduction to Science and the Scientific Method

The Layman's Guide to Making Sense of Statistics

John L. Campbell

VANTAGE PRESS
New York

Cover design by Susan Thomas

FIRST EDITION

Published by Vantage Press, Inc.
419 Park Ave. South, New York, NY 10016

Manufactured in the United States of America
ISBN: 978-0-533-16018-1

Library of Congress Catalog Card No.: 2008902301

0 9 8 7 6 5 4 3 2 1

In the furtherance of understanding

Contents

Preface

Nearly all books about statistics are textbooks for college students preparing to work in a field of endeavor requiring a professional's understanding of statistical concepts and techniques. Often these texts double as reference books consulted by individuals already in that field. Fields—professions—heavily into statistics include economics, business, psychology, sociology, education, public health, and medicine. The effects on society of applying to it the results of the statistical analyses performed daily in such fields run the gamut from shaping people's attitudes and determining the products and services they receive to molding public policy and evaluating ways of doing things in order to decide which are best. It therefore seems that everybody who isn't one of these professionals ought to have a practical knowledge about statistics and hence be statistically literate.

The title of this book indicates the book is for people ignorant of statistics. However, people fairly knowledgeable of statistics may find insights in it they can profitably apply. The book is intended to instill statistical literacy. Its concern with literacy makes it akin to my book intended to instill scientific literacy (Campbell, 2008). Probably most readers of this preface don't want to learn how to perform calculations with statistics and are only after the gist of statistics. This isn't surprising. After all, people today in need of these calculations are spared the drudgery which typically faced the people needing them thirty years ago. For today there are affordable personal computers that, when equipped with statistical software packages, process in seconds data inputted them, data which thirty or more years ago would take an hour or more to similarly process. Today, too, there are inexpensive handheld calculators that, while not as versatile as a PC, beat how statistical calculations once were done.

I recommend that you first skim this book, next read it thoroughly, and finally read it thoroughly again a month or two later. That way you will benefit from the pedagogical principle of distributed practice that facilitates the learning of tough topics. Should you later on wish to know how to perform statistical calculations you will need to consult a textbook, since you won't find in this book any formulae, equations, or exercises which pertain to calculating the statistics it discusses. Despite the book being introductory, it covers material in greater detail than many introductory statistics texts do. Hence don't be deceived by its lack of formulae, equations, and exercises. Now the reason behind its greater detail is the book was written with two audiences in mind.

Some readers want to learn statistics to know better the things heard or read which cite statistical information. Other readers need to learn statistics because knowing statistics is necessary or desirable in their line of work. Examples of readers of the second kind are individuals who routinely review statistical reports submitted to them and make important decisions or recommendations based upon what (they understand) these reports to say or not say. In writing this book I've tried heeding the advice of Albert Waugh (1943, p vii): "One of the most difficult problems in writing any elementary textbook is to keep it truly elementary—to write for beginning students rather than for one's colleagues."

The Layman's Guide to Making Sense of Statistics

1

Some Historical Perspective

Let's begin by clarifying what is meant by statistics. Statistics has three meanings. First, it means observations of something by someone. Such observations are popularly called facts and are subdivided into data and information.* They're referred to as data if either simple or relatively unorganized and as information if either complex or relatively organized. Hence statistics means observations, facts, data, or information concerning something, something observable and thus able to be observed or, more accurately, to be seen, heard, felt, smelled, etc. Statistics, so construed, takes the form of numbers, words, or mixtures of both. This first meaning of statistics arose around 1800 as the name for observations of characteristics of populations of people (note that populations of things other than people exist too). Most of the characteristics of human populations are of a socioeconomic—social or economic—nature. Labor statistics and vital statistics are observations which are ordinarily data and which address two major human population characteristics of the socioeconomic type.

Second, statistics means the profession that supplies other professions with concepts and techniques they need for processing data. Data processing is the transforming of less useful

* Note two things about the word data. One is that its first "a" is, phonetically speaking, hard and thus indicates data should be pronounced as spelled, as data and not as datta. The other thing is that despite people often using the word data to refer to a single item it is actually a plural word, the plural form of the seldom utilized singular word datum.

data into more useful information. Construed as a profession, statistics took shape about 1835 and about 1890 became a branch of mathematics. Third, since around 1950 statistics has meant the information in samples created from populations, a statistic being a specific type of information about samples or, in other words, a specific type of sample information. Statistics, construed as sample information, are usually mixtures of numbers and words. Now the context within which the word statistics occurs suggests which of its three meanings is intended.

Therefore "statistics" was coined around 1800 as the name for observations, typically data, that pertain to human population characteristics. Then, in about 1835 it came to also signify a profession furnishing other professions with the conceptual and technical tools they need to process their data and thereby make data into information. Finally, around 1950, statistics came to additionally mean sample information. It's important to understand that the advent of the word statistics did not signal the birth of collecting and processing data on human population characteristics. Those activities had, by 1800, been going on for a long time. Nevertheless its advent coincided with and was due in part to the mounting interest in socioeconomic data of capitalists thrilled with the newly unfolding Industrial Revolution and of social philosophers hoping to make into a reality the dreams associated with the Age of Enlightenment/Age of Reason (circa 1690–1790).

In what follows, we will sketch the history, up to approximately 1900, of collecting and processing data on human populations. To simplify matters, we will from here onward refer to said data as data, not statistics, and thereby limit our usage of the word statistics to its second and third meanings.

Curiosity and a need to know the characteristics of one's own population and of the populations of one's neighbors grow as the populations grow. And as these populations grow from dozens to hundreds to thousands to millions of individuals—from villages to towns to cities to nations—the observations (chiefly data) behind this knowledge tend to grow steadily in quality from the few and unplanned observations yielding what is known as anecdotal knowledge to the many and planned observations giving rise to systematized knowledge. It is the many, planned observations behind a systematized knowledge about populations that we'll be concentrating on. Consequently our journey into the past looks largely at populations which, in their day, were nations.

Need more than curiosity has inspired most studies designed to get population data. Hence a definitely practical air has mostly surrounded these studies. This should come as no surprise. It takes substantial time and effort to locate all, even most, inhabitants of a population, observe them, and record what's observed (i.e., record the observations made, typically seen as data collected). What needs could justify such vast outlays of time and energy? Many relate to a population's socioeconomic characteristics. Now by knowing a characteristic needing to be known, be it socioeconomic or otherwise, the ruler of a nation is able to rule more effectively. For instance, knowing the number of able-bodied men a nation has lets its ruler know how large an army his nation can have for defending itself or attacking a neighbor. Knowing their country has experienced over the last few years rising surpluses in its yearly harvesting of wheat and that a neighboring country's annual wheat harvest usually falls short of meeting its needs, wheat farmers of the first country know they have a potential market for their surpluses and hence a way to stop losing money (surpluses drive down prices, causing loss in revenue). Assuming that the neighboring country is friend and not foe, the first country's ruler has more than just one reason for getting together with its ruler to work out a trade deal.

Fundamental to wanting data on a population is appreciating the value of knowing this vs. (versus) that number of things, such as able-bodied men and bushels of surplus wheat. Without it, "numbers of things" mean nothing. Said appreciation existed at least 30,000 years ago and probably dates back farther than that. Evidence of it is seen in systematic notches cut 30,000 years ago into animal bones found here and there in Europe and the Middle East. They are interpreted as symbols constituting records of simple or relatively unorganized observations, and thus of data.

To this evidence may be added the thousands of carefully made clay objects stumbled upon for decades at archaeological sites in Iraq, Syria, Turkey, Israel, and Iran. They are fire hardened, as opposed to sun dried, and therefore wear resistant; are about half an inch to two inches in size; and were mass produced in seven standardized shapes from 8000 to 4400 B.C. and in nine more from 4400 to 3500 B.C. (Rudgley, 1999, pp 48–57). Their purpose was puzzling until the late 1980s, when the objects came to be viewed as the ingredients of a system to account for redistributing goods within a community. At that time they were aptly dubbed tokens and counters. It is thought their three dimensional shapes became the two dimensional shapes of the pictographs (picture symbols) on clay tablets made in Sumeria—southern Iraq—from 3500 to 3100 B.C. After 3100 B.C. the Sumerian tablets bearing pictographs were replaced by ones with wedge-shaped signs collectively called cuneiform, a writing which was popular in Iraq and its surroundings from 3100 to 100 B.C. Deciphering their pictographs and translating their cuneiform has revealed these tablets are predominantly records of business affairs: sales, purchases, receipts, stock inventories, deeds, etc.

Beginning approximately 1000 B.C. we see the start and gradual increase in rulers of European and Middle Eastern countries who order the systematic gathering of data on a population, ordinarily that of their nation, in order to come up with a systematized body of knowledge about that population. Like the notches on animal bones, the tokens, and the pictographs and cuneiform writing on clay tablets preceding them, the data obtained in these studies were tallies, counts, frequencies, or enumerations, and thus assertions of how frequently things of a certain kind (e.g., men, wheat) were observed. The inquiries typically aimed at observing *all* members of a population, such studies being complete enumerations termed *censuses*. Inquiries aiming at observing *some* population members, those picked to be members of a *sample* created to mirror—be a mirror image of—the population, are termed *surveys*. Therefore in population studies, surveys are to samples as censuses are to populations (surveys means something else insofar as parcels of land). While commonplace today, surveys of samples to learn, by inferring from their data, about the populations from which they had been created were largely distrusted and hence rarely done until the late 1800s. The Romans did 70 censuses between 435 B.C. and 35 (A.D.).

The nations in Europe and the Middle East continued to obtain data on populations through censuses after the fall of the (western branch of the) Roman Empire around 450, rise of the Islamic Empire about 650, and demise of the eastern Roman Empire (which commenced around 1800 to be known as the Byzantine Empire) in 1453. Although improvements in performing censuses no doubt came about during that thousand-year stretch, there does not appear to have been a strongly felt need for improvements. Then, in the Europe of the 1500s, that changed. The incentive for change was a critical stage reached in the political and economic development of the nations of western Europe, a development fueled by technological advances, social reforms, and the bourgeoning urge of a country and its citizens to realize their potentials.

Out of this ferment were born four distinctly European phenomena: the Protestant Reformation (circa 1525), the Scientific Revolution (circa 1600), the Age of Enlightenment/Age of Reason (circa 1700), and the Industrial Revolution (circa 1800).

Satisfying the needs modern nations have for data became an ever pressing challenge from 1600 onwards due to incessant changes taking place within them. This challenge forced countries to improve upon how they collect data and process data (convert it to information). What's more, it forced them to accept the fact censuses alone could no longer meet all their data needs! Besides a census commissioned every few years there must be ways to collect and process data at the time of the events the data signify happening. Recognizing this led to implementing methods to regularly collect and process data and to regularly send processed data—information—to those needing it. Some additional data management (data collection or processing) would be done by the government of a nation. The rest was left to the nation's merchants, bankers, manufacturers, etc., who needed much such supplementary data anyway to function properly.

About 1675 interest started being evinced in recording births, deaths, marriages, age at death, cause of death, and epidemics (their nature, date of onset, duration, locations, etc.). These records provided a basis in data—a database—for identifying trends and making predictions that would be beneficial to a country's functioning. A major impetus to compiling records was a book that Englishman William Petty wrote in 1679 about what he termed "political arithmetic." In it he offered insights into what to get data on and how to process data. His book started the ball rolling in collecting a rich variety of population data germane to a nation's functional integrity. The political arithmetic he conceived became a field of endeavor lasting roughly from 1700 to 1825. Petty's book was inspired by his friend and fellow countryman John Graunt, who in 1662 published tables summarizing as frequencies (counts) the voluminous unwieldy data contained in the English Bills of Mortality. As an aside, the words statistics and statistical were first employed in a treatise by Sir John Sinclair, another Englishman, entitled *The Statistical Account of Scotland (1791–1799)*. It was written around 1800. In it statistics is utilized in its original sense, to mean the observations of characteristics of populations of people.

Some of the best studies done in the 1700s to collect and process new or to reprocess old data on populations were done in England. And some of the best thinking on how to improve the way data are collected and processed was also English. However, notable studies and thinking along these lines took place elsewhere in Europe too, particularly in Germany. Two Germans stand out, Gottfried Achenwall and Johann Peter Süssmilch. In 1746 Achenwall started lecturing at Marburg on what he saw as useful vs. useless kinds of population data and on where to find existing records containing the useful kinds. Those attending these lectures spread his ideas throughout Europe. Süssmilch was an important advocate of care in selecting and using methods to calculate annuities and life insurance, a couple of relatively recent innovations, employing mortality-type population data. In the late 1700s there commenced springing up in the more progressive nations of Europe government bureaus whose staff maintained records of population data they or others in a country regularly collected or processed. Such "statistics bureaus", however, did not become fairly commonplace until the mid 1800s.

Approximately 1800, there started occurring in much of Europe, followed by America, a transition from national economies based primarily on rural and agricultural ways of making a living to national economies based chiefly on urban and industrial livelihoods. Coinciding with the transition was an upsurge in capitalism, reorientation in banking, and mushrooming of commerce (trade) between nations. The newly industrialized countries quickly found they had an even greater need than before to regularly acquire data so they could adequately monitor and thus know the states of their now more intricate economies. Naturally most of these data addressed industry and commerce. Since data on economics imply the well-being of a nation and since "nation" is, in Greek, "polis", economics data were soon considered, in an extended sense, political (polis) data.

Statistics, as a field of endeavor, has two origins. One is the political arithmetic of William Petty just alluded to. The other is anthropometry. Anthropometry collects and processes anthropometric data, data on physical characteristics of humans (e.g., average arm length, average skull capacity). Belgian mathematician and astronomer Lambert Adolphe Jacques Quetelet (pronounced "ketlay") became interested in anthropometry about 1825. During his spare time he measured and compared the chest size of Scottish soldiers, height of French army officers, etc. and in so doing pioneered anthropometry. Of interest here is he discovered and was fascinated with regularities in anthropometric measurements (data). One such regularity was the precise way—precisely how—they varied or were dispersed. Specifically, they regularly varied the same way astronomical observations had lately been seen to regularly vary, a way we today refer to as either the normal distribution or the bell curve. Normal distribution and bell curve are descriptive labels. Normal connotes regular: the way data vary that's termed normal is the normal way data vary or the way data normally vary. Bell curve connotes that if data are plotted as points on graph paper and if adjacent plotted points are connected by drawing lines between them then these lines become a single line which is not only curved as opposed to straight but, with some imagination, looks like an old-fashioned church bell.

Around 1835 the approach to data that had characterized the by then defunct political arithmetic merged with the approach to data being taken by Quetelet in anthropometry to produce the field of statistics. From political arithmetic, statistics acquired the practice of utilizing arithmetic (addition, subtraction, multiplication, division, and making ratios) to process data, make data into information. From Quetelet, statistics acquired the practice of employing distributions conceived by a branch of mathematics known as probability theory to more fully process data and to make reasonable inferences about populations utilizing data from samples created from those populations. The normal distribution of data which so fascinated Quetelet was the beginning of this distribution being, out of all distributions embodying probability theory, the most useful to statistics. Yet a few others would prove useful too, especially one called the binomial distribution. Incidentally, the distributions of probability theory are referred to equally as probability distributions and frequency distributions, since they address both probabilities (the chances such-and-such will happen) and frequencies (how often such-and-such did happen).

The relationship between statistics and probability theory initiated by Quetelet's awe of how anthropometric, astronomical, and even other types of data varied in a fashion that resembled the shape of the normal distribution grew stronger as the years passed. In fact the two fields became as if they were two peas in a pod after statistics joined probability theory as a branch of mathematics around 1890. Consequently it would be good for us to take a look at probability theory.

Probability theory became a branch of mathematics about 1655. Its birth was not due to interest in knowing the characteristics of human populations. Rather, it was due to an interest in gambling which spread like wildfire across much of Europe in the 1500s and 1600s. Gambling was particularly popular in

France and with aristocrats, who wanting to get richer wanted to know as accurately as possible their chances of winning at games of chance. During the 1500s the chances, or chance, of an event occurring were customarily stated as the odds in favor of, or favoring, its occurrence on an occasion, say on a spin of a roulette wheel. In the 1600s the chances of an event occurring came to be stated also as the probability, or likelihood, of its occurrence, some people finding it easier to interpret and apply to situations a probability (one number) than odds (two numbers). Chances are still expressed as odds, mainly in gambling situations.

Attempts to pin down one's chances in games of chance led to realizing chance is not unpredictability, only much less predictability than desired. Out of this realization came probability theory, a branch of mathematics devoted to discovering and describing the regularities, orderliness, or lawfulness in chance happenings. The theory gave gamblers and others (e.g., insurance companies) ways to know and apply the laws of chance or, as it's also called, randomness. Probability theory stemmed from an exchange of letters in 1654 between two Frenchmen, the mathematician Pierre de Fermat and mathematician-physicist Blaise Pascal. In their letters they pieced together the first of various laws of chance, laws of probability, or rules of probability. Other laws would follow. One was set forth in 1705 by the Swiss mathematician Jacques Bernoulli. It allows us to state the probability of an event occurring in situations that are characterized as Bernoulli trials or a Bernoulli process. Such a probability is found by consulting a probability distribution Bernoulli conceived, the binomial distribution. This distribution was the first of many probability distributions.

We'll end our digression into probability theory with the normal distribution. Unlike the binomial distribution, the normal distribution was conceived over a roughly eighty-year period and by three people: the French mathematician Abraham de Moivre (1733), French astronomer-mathematician Pierre Laplace (1778), and German mathematician-astronomer Carl Friedrich Gauss (1809). Now astronomers and physicists during the 1700s increasingly assumed that the things they observed and the relationships between them were regular occurrences which implied nature or, what came to mean the same thing, the universe operates much like a viable society does, by being orderly or lawful (law abiding) and thus by exhibiting lawfulness. Irregularities, deviations, or dispersion were noticed. However, they were fairly small and were often passed off as individual differences among observers (e.g., this vs. that scientist's eyesight, equipment, expertise) and among observations the same observer made at different times. Seldom were they ascribed to nature. By 1800 the irregularities, deviations, or dispersion from an assumed normal state of existence, the norm, for a thing or a relationship between things had come to be interpreted by astronomers as errors—mistakes made—in observing.

Gauss thought the probability distribution he helped conceive was a mirror that reflected the form of precisely how the observations of most things are distributed and therefore how the deviations or dispersion of these observations can be precisely measured. In keeping with his construal of deviation as error, he named the distribution the law of error, or error law, and error curve. During the late 1800s it was renamed the Gaussian law (of error), Gaussian curve, and Gaussian distribution. In the early 1900s the distribution came to also be called the normal law (of error). Since around 1950 it usually has gone by the names normal distribution, normal curve, and, in everyday talk, bell curve. Gauss' fascination with how well the distribution described dispersion in observations would fascinate Quetelet too, twenty years later.

Social reform was in the wind in Europe and America during the mid and late 1800s. Into this wind was swept the newborn field of statistics, largely because of claims made by Quetelet. Around 1845 Quetelet said not only anthropometric-type data but all types of data on people are normally distributed. In order to know exactly what he, and Gauss before him, meant by normally distributed data and a normal distribution of data we must take a moment to learn exactly what they meant by norm. For a norm was their basis for judging the existence and extent of irregularities or deviations in observations (data). Of course it was actually the sum total of all deviations in observations that really interested them. This sum total is termed the dispersion in or variability of observations. Now, to them norm meant average. Hence dispersion meant all deviations in observations from, or to either side of, the average observation. Average meant the average of a collection or set of data. And it meant the average of the frequency distribution of a set of data, a frequency distribution being the distribution of the frequencies (counts) of the different quantities/categories (data) in the set. Both meanings of average are equivalent, since the average of a set of data is the average of the data set's frequency distribution.

More controversial than his claiming *all* types of human data are normally distributed was Quetelet's view on what deviations (errors) or dispersion (error) imply. Gauss saw them as imperfections in observing things and thus as imperfect observations. Quetelet saw them as imperfections in things observed and thus as imperfect things. Therefore a thing (e.g., a person) is perfect with respect to some characteristic of it (e.g., height) if the thing does not deviate from the average of this characteristic and is imperfect if it does. Suppose the average height of adult men in America is 5′8″. Quetelet would say 5′8″ is the perfect height for adult American men and that all adult American men 5′8″ tall are perfect, in height. All shorter or taller than 5′8″ would be imperfect, in height. A person perfect in all respects—height, weight, health, education, income, morality, etc.—is the perfect person and the perfect person is for Quetelet the average person or, as he put it, the "average man".

We've illustrated perfection and average with the anthropometric characteristic height. But by 1845 Quetelet's interest in perfection and average was directed at socioeconomic characteristics. Why? Because he apparently hoped that nations would adopt them as norms in social reforms for elevating to average those of their citizens below average in health, education, income, etc. Quetelet's ideas were provocative, provoking debate on at least four fronts. One front was the pros and cons over government intervention in people's lives. Another was fear that economic disincentives would result from averaging out, or homogenizing, personal incomes. Still another front was the wisdom of concentrating on an average to the exclusion of the dispersion associated with it. Finally, a few thought much could be learned about people by concentrating on dispersion to the exclusion of its associated average, a stance which led during the late 1800s and early 1900s to the study in biology and psychology of individual differences.

Amidst the hubbub about its social implications, statistics continued doing what political arithmetic had done before it; namely, collect and process population data important to a country's functioning. From 1850 onward statistics evolved into a key tool and aspect of government, economics, and business. Breakthroughs came in making tables and graphs more informative, graphing trends, and creating price index numbers. In the late 1800s statistics started to be applied to areas within biology, initially heredity. Around 1890 it became a branch of mathematics, bringing it closer to probability theory. Shortly after that, statistics began refining its concepts of average, dispersion, relationship, population, and sample. Also it soon began to draw heavily upon probability theory in developing concepts and techniques for use in making reasonable inferences about populations from the data in samples created from them. Early pioneers of a more mathematical statistics (1880–1910) were English: Francis Edgeworth, Francis Galton, Karl Pearson, Udny Yule, Charles Spearman, William Gossett.

From its inception as a profession around 1835 until roughly 1910 statistics lacked a clear identity. The reason is that its identity was mainly the identity of the profession of which it was an aspect and for which it was the provider of tools meeting needs specific to the profession, be this profession economics, biology, psychology, education, etc. Thus a statistician was often an economist (Edgeworth), biologist (Galton), psychologist (Spearman), or even industrial chemist (Gossett) interested in applying statistics to his field of endeavor. An English economist, William Bowley, finished writing in 1901 a very early textbook on statistics, one having an economics slant. Frequently a profession regarded statistics as an important element in a still larger aspect of it, an aspect that was a specialty concerned with data collection or processing. Instances of these specialties are econometrics in economics, biometrics in biology, and psychometrics in psychology. Only after about 1910 did statistics' identity become clearer.

2

The Nature of Statistics

The last chapter gave a preliminary impression of what statistics is all about. In this chapter we will get a better impression. Let us begin by taking a look at the purpose of statistics. Its purpose is to tell us about a group of things (e.g., men) or a relationship between groups of things (e.g., men and women). Therefore the purpose of statistics is not to tell us about a single thing (e.g., Paul) or a relationship between things (e.g., Paul and Carol). Probability theory has the same purpose that statistics has. Now it's their having a shared purpose that mainly led us on page 9 to liken statistics and probability theory to two peas in a pod. This commonality is partly the reason we sometimes come across a textbook on both probability (theory) and statistics.

Consequently this book deals with situations where interest is in a group of things, not in a thing of a group, and in a relationship between groups of things, not in a relationship between a thing in one group and a thing in another group. This is not to say we will be ignoring these things. Far from it. We'll be up to our eyeteeth in them. However, our interest in things will be limited to treating them as means to an end. The end is to know a group of things or a relationship between groups of things. And the means of attaining that end is knowing the things in the group or, for a relationship between groups, the things in two or more groups.

Before we leave the subject of distinguishing between knowing a group of things and knowing a thing it might be smart to firm up the implications of this distinction with an illustration. Consider the group of Scottish soldiers whose chest size Quetelet measured and compared (page 8). We do not know how many he studied. Assume that he only had time to study 200. A group so small would be a sample, not the population, of Scottish soldiers. Assume, too, upon their being listed from the smallest to the largest the 200 measurements are observed to go from 36 inches through 48 inches.* If interest is in the group of soldiers, interest is in all 200 chest measurements and in utilizing them to find the group's average chest size and dispersion in chest size. Suppose the measurements are normally distributed. This means they will produce an average of 42 inches and may produce a dispersion of, say, three inches to either side of 42 inches (in 68 percent of the 200 soldiers). Now if the interest is in a soldier instead of the group then interest is in one chest measurement, a measurement useful to know when outfitting that soldier yet, by itself, useless to know when the aim is to know the group.

Implicitly, we're treating the above distribution as a frequency distribution. Thus it gives us the frequency, count, or number of times the 200 measurements fell into one or the other of the perhaps thirteen chest sizes into which we divvied them up in order to obtain a panoramic view, the big picture, of their distribution. The chest sizes might be 36 inches, 37 inches, 38 inches,—46 inches, 47 inches, and 48 inches; the frequencies associated with them might be 4, 11, 18,—19, 11, and 2. We can treat this frequency distribution as a probability distribution. By doing so we can know our probability or chances of, say, meeting in the near future a soldier from the

* Note that most data are measurements. A measurement is a number (e.g., 36) followed by a unit of measurement (e.g., inches).

group, given we do meet one, having a chest measurement of such-and-such. This probability, however, will be accurate only if whatever soldier we meet is met on a chance or random basis. Therefore the soldier we meet, or simply observe, must be a random or chance event in order for us to accurately know the probability of occurrence of a soldier—of our meeting or simply observing a soldier—from the group having such-and-such chest size. Now for a word of caution. To be assured under everyday circumstances that the occurrence of a thing is a truly random occurrence is not as easy as it may seem!

Statistics is interested in two types of groups, termed populations and samples. A population is a group of all things of a certain kind (e.g., all Scottish soldiers) while a sample is a group of some things of a certain kind (e.g., some Scottish soldiers). Thus samples reside inside populations. Many samples, each differing from all others in its things, can be created from a population. Statistics is primarily interested in populations. However, almost never can we observe all things of a certain kind (a population). Nearly always we must be content to observe some things of a certain kind (a sample). Just as knowing things is the means to knowing a sample, knowing a sample is the means to knowing a population. Consequently, out of necessity, statistics is secondarily interested in samples. If the things in a population which are selected to be the things in a sample resemble things not selected from the population, the sample created is variously said to represent, to mirror, to reflect, or to be representative of this population.

The process of creating, obtaining, drawing, piecing together, etc. a sample from a population is called sampling. A population subjected to it is said to be sampled and is called the population sampled. Sampling is widely utilized in surveys and experiments. Therefore we need to spend some time discussing it. We begin with an illustration. It involves a nonhuman population, to illustrate our point on page 1 that populations of things other than people also exist.

Consider the Whitman Sampler. The Whitman Sampler is a box of chocolate candies manufactured by Whitman. It gets the Sampler in its name from being a sample of the chocolates Whitman makes. Actually the Whitman Sampler is many samples, because many Samplers have been packaged over the years. Samplers come in small, medium, and large boxes. A small box contains roughly a dozen chocolates, a medium-sized box about two dozen, and a large box approximately three dozen. The chocolates in every Sampler, irrespective of its size, are selected with the hope that they reflect the assortment in chocolates the company makes during a particular span of time. In other words, every Sampler is so constituted as to represent the entire line of Whitman's chocolates (during a particular time span), an entirety which is the population of Whitman chocolates (during that time span).

The logic underlying the Sampler is we will think or, more precisely, infer that the population of chocolates is as good as we found the sample of chocolates, the Sampler, to be. Our rating of the goodness of the Sampler arises from our observing and thus seeing, tasting, smelling, and touching its chocolates. The ensuing observations take the form of data on the varying amounts of pleasure we attribute to a chocolate's color, shape, size, taste, firmness, etc. From data we collect on the chocolates in our sample of Whitman chocolates we generate information on and thereby knowledge about chocolates in the population of Whitman chocolates. All this we can do in our heads without aid of pencil and paper, calculator, or statistics. If we want a more accurate knowledge than this, we must observe more than one Sampler.

Let us build upon our illustration so we can learn more about the nature of statistics. Suppose we want a more accurate knowledge about the above population. This means we need to follow a procedure. It has four steps. First, we create a random sample which contains as things a minimum of ten

Samplers. The randomness of this sample assures us an essentially representative sample and a sample which satisfies an assumption made by most techniques using samples to tell us about populations. Second, we collect data from the sample. Our hope at this point is that the sample information (a known) generated in the third step is, in the fourth, judged to be essentially the same as population information (an unknown). Third, we process the sample data collected, transforming it into sample information. Fourth, we decide whether the sample information is vs. isn't essentially the same as the population information.

Our example of sampling is an example of research too! It exemplifies the sort of research done by companies to learn what customers, actual and potential, think of a product they manufacture. Specifically, it's an instance of the sort of taste preference research done in fancy laboratories by big food companies such as Wesson Foods. Of course our illustration is a much watered down version of what happens inside product evaluation labs. Were Whitman to do a study on the appeal of its chocolates, it would first solicit more than one person's opinion. It would probably draw, on a random basis, a sample of a minimum of fifty people from the population of people liking chocolates. Then it would assign (give), at random, one Sampler to each of them. Next it would ask each to eat the chocolates in his or her Sampler and rate how much he/she enjoyed them (we won't get bogged down with the numerous details on method that a researcher must face here, such as how many chocolates are eaten over how long an interval of time and controlling the bias caused by aftertaste or by which chocolate a person ate first). Finally the ratings—data—would be processed into information on the population of Whitman's chocolate candies.

The Sampler(s) chosen in the last three paragraphs may be small, medium-sized, or large. Now the more chocolates in a Sampler and thus a sample, the more accurate the knowledge about the population of chocolates. And the more people in a sample, the more accurate the knowledge about the population of chocolates. Consequently in all three foregoing scenarios a large Sampler is preferable to a medium-sized and especially a small Sampler. In the second scenario fifteen Samplers are preferable to the minimum of ten. And in the third scenario a sample of sixty people is preferable to the minimum of fifty. Not that quantity is a substitute for quality! If in the first scenario the sole Sampler isn't randomly chosen and randomly constituted, we're apt to get a mistaken impression of the population of chocolates regardless of that Sampler's size.

We need to look at the reasoning behind employing samples to know populations. A sample has far fewer things in it than does the population yielding it. In fact a sample typically has one percent or less things in it compared to its parent population. How can knowing so few things result in knowing so many? It is possible if a sample can be said to represent a population. Strictly speaking, a population may be known by knowing a sample if the sample is *essentially* representative of the population. For then sample and population information are *essentially* the same. By essentially representative is meant a sample is indistinguishable from a population in characteristics and in the states in which characteristics exist. Essentially the same as means that differences between sample and population information stem only from the presence of random influences (the operation of chance) and not from any systematic influences whatsoever. In statistics, differences due to random influences are regarded as not significant and differences due to systematic influences are considered significant. Significant differences are noteworthy.

There are seven basic types of samples used in acquiring information on populations. Each is named after its key characteristic. Hence the convenience sample is convenient to come by, to assemble. Its things are rounded up fairly easily and selected based upon judging them to collectively be representative of the population of interest. Convenience samples are also known as judgment samples. A convenience sample, carefully pieced together, can usually be legitimately considered a representative sample. A presenting sample is relatively easy too, since things chosen for it present themselves to the person doing the sampling on their way to some destination (e.g., moviegoers walking through a theater door or partially assembled lawnmowers on top of a conveyor belt passing by an inspection point while headed for the next assembly point). The representativeness of a presenting sample is enhanced by selecting every, say, fifth or fifteenth thing that comes along as opposed to every thing or every other thing.

A systematic sample is like a presenting sample. Here a list of things in a population exists and is utilized to systematically construct a sample, systematic being connoted by repetitively picking every, say, fifth or fifteenth thing listed for inclusion in the sample. Such a list is technically termed a sampling frame, which means a frame of reference as to things constituting the population sampled. Often the individual creating a systematic sample begins by choosing at random where on the list, or frame, to start picking things, a random start being often thought to further enhance sample representativeness. Hence the expression systematic sample with random start. A (simple) random sample is one out of all possible samples of a certain size, for instance twenty things, that had the same probability of becoming the sample obtained. Inasmuch as all these same-sized potential samples had equal probabilities of being the actual sample, the sample picked is the result of random influences. Most statistical techniques that use samples to acquire population information require random samples. The same techniques require random groups when an experiment employs groups created from a sample, a random group being created by randomly assigning things to it from the sample.

A stratified systematic sample, popularly referred to as a quota sample, is a composite systematic sample, meaning two or more systematic samples united after being created. The two or more component systematic samples reflect strata in a population of interest, usually a human population. Each component systematic sample represents a population stratum. Population strata are populations within a population. They are subpopulations. Population strata—subpopulations—are acknowledged because the things in them have a characteristic (e.g., gender) whose states or modes of expression (e.g., male and female) are hard to represent accurately in a systematic sample, hard insofar as the percentages of these states in the population. A characteristic viewed as stratified is dubbed a stratified characteristic. Now a stratified characteristic is not the characteristic of primary interest when assembling a stratified systematic sample. It is of secondary interest though. For if the percentages of its states aren't well mirrored by a sample then there will likely arise a distorted impression of the characteristic of primary interest (e.g., wages).

A stratified random sample resembles a stratified systematic sample. The difference between them is that with a stratified random sample its two or more component samples are random samples, not systematic samples. Other than this, a stratified random sample is the same as a stratified systematic sample. The last basic type of sample is a cluster sample. It's almost exclusively employed for getting information on human populations. In sampling, clusters are customarily regarded as being geographic things in a population whose identities are readily agreed on (e.g., city blocks, county lines, census tracts, map grid coordinates). Clusters are basically subpopulations. Usually each of the numerous (e.g., 100) clusters into which a population being sampled has been subdivided is a tiny percentage (e.g., one percent) of that population's people. A cluster sample is a sample whose things are clusters of people randomly picked from a frame listing all population clusters. Thus a cluster sample is a random sample whose things are clusters (of people) rather than people.

We end our discussion of sampling with three remarks. First, the types of samples most likely to be (essentially) representative are those whose composition was clearly determined by randomness, or chance. This is why random samples, stratified random samples, and cluster samples are appealing to researchers. Second, a few of the many techniques assuming data they process come from random samples have been modified so they can be employed with systematic, stratified random, and cluster samples. Third, the procedures, methods, or plans used to draw the aforementioned samples go by the names convenience (judgmental) sampling, presenting sampling, systematic sampling, (simple) random sampling, stratified systematic (quota) sampling, stratified random sampling, and cluster sampling.

It is time to look in detail at how statistics allows us to know a population by knowing a sample obtained from it. Put another way, our attention will be directed at how we can know whether sample information and population information are essentially the same. Let us begin by stressing we can never know whether sample information is vs. isn't the same as population information. For while sample information becomes known through processing the sample data collected, population information remains forever unknown. All we can ever aspire to know is whether the two are vs aren't essentially the same or, equivalently, whether the difference between the two is vs. isn't due to random (chance) influences and therefore isn't vs. is significant. To know this, we must adopt a criterion, a dividing line, which splits all conceivable differences between the two into those we will regard as small enough to be probably caused by random influences vs. large enough to be probably caused by systematic influences.

Sometimes we are more interested in knowing if sample and population information are essentially the same than knowing if they are significantly different. In such cases we're also more interested knowing our confidence with respect to being right saying that the two are essentially the same than knowing our risk with respect to being wrong deciding they are significantly different. Such interests arise when we want to use sample information to estimate population information. Usually we are more interested knowing if sample and population information are significantly different than knowing if they are essentially the same. In such cases we're more interested knowing our risk in being wrong deciding they are significantly different than knowing our confidence being right saying they are essentially the same. Such interests arise when we want to show that sample information is not from the population which at the outset of a study is hypothesized to be the source of data the study will collect and process into sample information.

Now the criterion we adopt will in its specifics be governed by the level of confidence or of risk we seek. The more confident we wish to be, the higher will be our criterion as to what is small enough to be probably caused by random influences vs. large enough to be probably caused by systematic influences. Similarly, the less risk we wish to take, the higher will be our criterion on what is small enough to be probably caused by random influences vs. large enough to be probably caused by systematic influences. Therefore the criterion varies as our desired level of confidence/risk varies. Both confidence and risk are stated as percentages (e.g., 95 percent or 5 percent) linked to the probabilities (e.g., 0.95 or 0.05) found in probability distributions. Their percentages can never be 0 or 100 because 0 percent and 100 percent signify certainties and never in these matters can we be certain of anything! Hence we come face to face with an important truism about using samples to know populations; namely, there is a distinction between reality (e.g., sample and population information are essentially the same in such-and-such situation) and what we believe reality to be (e.g., we are 95 percent confident saying sample and population information are essentially the same in said situation).

An analogy may help solidify the distinction statisticians make between systematic variation and random variation. That distinction is analogous to the distinction made in signal detection situations between signals and noise. Consider a situation in which the signal to be detected is a faint sound and the noise is a slight, perpetual ringing in your ears. Whenever the signal occurs, it is embedded in and superimposed upon the noise. Now you're deciding whether an interval of time contains the signal and the noise vs. the noise is akin to a researcher deciding whether the data in her study exhibit systematic variation immersed in random variation vs. random variation.

Statistics engages in two basic activities. One is describing samples and populations; the other is using samples to make inferences about populations. Statistics is therefore subdivided into two areas, descriptive and inferential statistics. Descriptive statistics is concerned with concepts and techniques that address transforming group data into group information and consequently sample data into sample information and population data into population information. Naturally most of its concern is with processing sample data into sample information, since seldom do we have population data (i.e., data on all things in a population) at our disposal and thereby the ability to process population data into population information. Inferential statistics is concerned with concepts and techniques that address transforming sample information into population information. Processing sample information into population information entails inferring that sample information and population information are, depending on what interests us, essentially the same or significantly different. Thus inferential statistics is also concerned with specifying the criterion we adopt for saying/deciding sample and population information are vs. aren't essentially the same/significantly different and with the confidence/risk level that affects this criterion. Note that although inferential statistics deals with inferences it, in effect, deals with descriptions too. For the inferences it makes describe populations and hence are inferred descriptions of populations.

Inferential statistics has as its main interests the two interests that its users normally have. Therefore it is mainly interested in estimating population information and, especially, in determining the significance of differences. Inferential statistics commonly employs a couple of terms we should introduce. They are the technical equivalents to what we've been terming sample information and population information. Sample information and population information are respectively called statistics and parameters. Hence the two main interests of inferential statistics are estimating parameters—producing parameter estimates—and determining significances of differences regarding statistics.

The final thing needed in learning the nature of statistics is knowing what statistics means by information. What is meant by sample information (statistics) and population information (parameters)? As used in statistics, information is hard to define in a way that is meaningful. Back on page 1 we said information is either complex or relatively organized observations. However, that is the definition of information in its usual sense and isn't quite what statistics has in mind. And to say that in statistics information, which is more accurately termed statistical information, is what is generated from sample data or, though nearly never, from population data leaves much to be desired. So we will not attempt a definition and instead will identify the types of such information.

Being that sample information and population information are descriptions of, and thus describe, samples and populations in terms of their characteristics, a little time ought to be spent understanding these characteristics before embarking on identifying the types of statistical information. To understand the characteristics of samples and populations is to realize they are also the characteristics of things and is to know how characteristics of things are classified. We'll examine two classifications of characteristics.

The oldest and best known classification splits characteristics into qualitative and quantitative characteristics, qualitative characteristics being increasingly referred to as categorical characteristics since the 1960s. A categorical/qualitative characteristic (e.g., ethnicity, or ethnic background) manifests itself in states we think of as categories which differ in kind (e.g., Greek, Swedish, Angolan, Japanese, Mexican, Russian, and Iranian); a quantitative characteristic (e.g., temperature) exhibits states considered quantities (e.g., 50 degrees and 75 degrees Fahrenheit). A refinement made in this classification during the late 1940s is fairly popular with psychologists and sociologists. It calls a categorical characteristic a nominal characteristic and calls a quantitative characteristic an ordinal, interval, or ratio characteristic, depending on how its quantities differ.

An ordinal characteristic (e.g., intelligence) has quantities we can distinguish between only by saying one of them (e.g., an intelligence quotient—IQ—score of 120) is less than or greater than another (e.g., 125). Many psychological characteristics are ordinal, even though some may seem interval due to how they are defined. An interval characteristic has quantities we can distinguish between by additionally saying one of them is less than or greater than another by such-and-such. And a ratio characteristic (e.g., weight) is an interval characteristic having a complete absence of quantity which is symbolized by the number zero. Interval characteristics are mostly conveniences, created so we can more easily deal with states of certain ratio characteristics. For instance, temperature is a ratio characteristic ordinarily treated as an interval characteristic. While the scales in terms of which temperature is commonly measured (i.e., the Fahrenheit scale and the Celsius, or centigrade, scale) contain zeros (i.e., 0 degree Fahrenheit and 0 degree Celsius) the zeros don't signify no temperature! Envisioned as an interval characteristic, temperature has quantities (temperatures) we can distinguish between by saying one of them (e.g., 50 degrees Fahrenheit) is less than or greater than another (e.g., 75 degrees) by such-and-such (25 degrees).

Eight types of statistical information are commonly employed in statistics. They are identified and defined below:

1. Frequencies: A frequency, or count, is the number of times a particular state (e.g., male) of a particular characteristic (i.e., sex) of a group (e.g., historians living in Texas) was observed (e.g., 10). By a group is meant, here and in what follows, a sample or a population. Also a frequency is the number of times (e.g., 5) an event (e.g., "false-positive") associated with a group (e.g., the outcomes of administering a certain medical test) was observed, event being a term often encountered in probability theory and meaning the *occurrence* of a state of a categorical, or nominal, characteristic.

2. Frequency Distributions: A frequency distribution is a portrayal of frequencies of states of a particular characteristic (e.g., sex) of a group (e.g., historians living in Texas), each frequency referring to a particular state (e.g., male) of that characteristic. If the characteristic is quantitative, its frequencies are arrayed and hence made into an array. By an array means that the quantities of a quantitative characteristic (an ordinal, interval, or ratio characteristic) are arranged into a series that, when looked at from left to right, either steadily increases in size (e.g., 34, 38, 41, 49, 55, 70) or steadily decreases in size (e.g., 70, 55, 49, 41, 38, 34). A frequency distribution also is a portrayal of the frequencies of events (e.g., false-positive, false-negative, true-positive, true-negative) associated with a group (e.g., the outcomes of administering a certain medical test). Frequency distributions can take the form of a table of frequencies—a frequency table—or the form of a graph, where frequencies are depicted either as bars (i.e., a bar graph or a histogram) or as lines drawn between plotted points (i.e., a line graph that here is more precisely termed a frequency polygon, since its shape is that of a polygon).

3. Relative Frequencies: A relative frequency (e.g., 0.10), a commonly encountered kind of proportion, is a frequency (e.g., 10) of a state that's been made relative to, or thus been divided by, all frequencies (e.g., 100) of states of a particular characteristic (e.g., sex) of a group (e.g., historians living in Texas). Also a relative frequency (e.g., 0.02) is the frequency (e.g., 5) of an event (e.g., false-positive) which is relative to all frequencies (e.g., 250) of events (e.g., false-positive, false-negative, true-positive, true-negative) associated with a group (e.g., outcomes of administering a certain medical test). Relative frequencies of states of a particular characteristic of a group or of events—occurrences—associated with a group can, like frequencies, be portrayed via a frequency distribution. When this is done, the frequency distribution is sometimes called a relative frequency distribution. Now, as for employing the words proportion and relative frequency, there are times one is preferred over the other. We'll use both words.

Notice that a relative frequency provides an often desirable perspective on the frequency underlying it by showing where that frequency stands *in relation to* all the frequencies constituting the frequency distribution for the group being considered. For instance, the relative frequency 0.02 indicates that the frequency underlying it (which may be any one out of many frequencies, but above was 5) is very small, compared to all frequencies. Notice, too, that an easier to grasp number results from multiplying a relative frequency by 100, thus converting it into a percentage. Hence, by multiplying the relative frequency 0.02 by 100, we get the percentage 2%. However, doing this when the sum of all frequencies for a group totals less than fifty is deceptive since *percent* or, more accurately, *per cent* means *per one hundred* and consequently signifies a basis—a reference point—of one hundred separate numbers (Blalock, 1979, pp 33–34). Which is why it's a good practice to give along with a percentage the sum of the frequencies, the 'base', used for calculating its associated proportion (250, in the above example). My advice for what to do when you see a percentage is this. If the base cited for it is much less than fifty (e.g., forty), be wary of what it says it is indicating. And if its base is not cited then ignore it.

4. Averages: An average is the typical state of a frequency distribution or, in other words, the typical state exhibited by a characteristic of a group. Also an average is the typical event of a frequency distribution. Statistically-minded people speak of central tendency instead of average, though we'll use the more popular word average. Average, or central tendency, can be expressed in various ways. Each of these ways is regarded a measure of average (central tendency). The major measures of average are the arithmetic mean and the median. Minor measures include the weighted arithmetic mean, the mid-range, the mode, the harmonic mean, and the geometric mean.

A couple of things should be noted. First, the arithmetic mean is, by far, the most commonly utilized average. Therefore it's often called in everyday conversation and even in technical discussions the average (e.g., the average annual rainfall in Boston is such-and-such) rather than the mean and the mean rather than the arithmetic mean. This is fine, as long as individuals coming across the terms average and mean know exactly what's meant by them. Second, the different kinds of averages exist because each portrays average—typicality—better under certain circumstances than under others. Examples of such circumstances are whether a characteristic is interval/ratio versus ordinal, whether a characteristic's frequency distribution is mound-shaped versus lopsided, and whether in a specific situation some states of an interval/ratio characteristic ought to exert more of an influence or carry more weight than others in determining average in that situation.

5. Dispersions: A dispersion, variability, scatter, or spread is the extent of variation or change in states displayed by a characteristic of a group. Also a dispersion is the extent of variation in events of a frequency distribution. Dispersion, like average, can be expressed in various ways. Each of these ways is regarded a measure of dispersion. The major measures of dispersion are the standard deviation, the variance, the quartile deviation (semi-interquartile range), and the coefficient of variation. Minor measures include the range, the average deviation, the variation ratio (Blalock, 1979, p 76), and the index of qualitative variation (Mendenhall et al., 1974, p 121). Seldom are dispersions of interest in their own right. Rather, interest in dispersions is ordinarily for employing them to gauge how much stock we ought, or ought not, to place in the averages associated with them. Therefore a dispersion bestows meaning on and thus establishes the usefulness of an average: the more dispersion in states of a characteristic, the less meaningful and useful the average state of that characteristic!

6. Differences: The commonly sought differences are those between frequencies, between proportions, between sets of ranks, between averages (mainly arithmetic means), and between dispersions (mainly variances). Ranks are numbers assigned quantities manifested either by an ordinal characteristic or by an interval/ratio characteristic treated as an ordinal characteristic. Thus no more can sensibly be said about two different ranks—ranked quantities—than that one of them is less than or greater than the other. A set of ranks normally has its constituent ranks ordered (arrayed), usually from smallest to largest. The term rank-order signifies a set of ranks whose constituent ranks have been ordered (arrayed). Differences between two or more sets of ranks are, to be more precise, differences between two or more sets of ordered ranks or, in other words, differences between two or more rank-orders.

Most research studies are designed around calculating differences between frequencies, relative frequencies, sets of ranks, averages, or dispersions in the hope of discovering relationships between characteristics. Those of these studies which are experiments specifically hope to discover causal relationships between them.

7. Correlations: A correlation, also known as an association, is the *strength* of the relationship between states exhibited by one characteristic (e.g., height) and states displayed by one or more other characteristics (e.g., weight) of a group (e.g., U.S. adults). The reason for italicizing the word strength is to stress correlation is not just another word for relationship, as all too often is indicated (even in textbooks). Rather, correlation is strength of relationship. The term implies the strength of a relationship ranges anywhere from very, very weak (essentially no relationship) to very, very strong (essentially a perfect relationship).

Correlation can be expressed in various ways. Each way is regarded a measure of correlation and is usually termed a coefficient of correlation, or correlation coefficient. The major measures of correlation are the Pearson product-moment correlation coefficient (occasionally called "the correlation coefficient", which breeds confusion) and the Spearman rank-order correlation coefficient. Minor measures include the Cramér coefficient, the correlation ratio (also referred to as "eta"), the point-biserial correlation coefficient, the Pearson product-moment multiple correlation coefficient, the Pearson product-moment partial correlation coefficient, and Goodman and Kruskal's gamma. Besides these measures of correlation, which specify correlation by a number (typically, from 0.00 through 1.00 and indicating no through perfect correlation), correlation can also be seen in the density and the configuration of points plotted on graph paper and forming a graph termed a scatter diagram, scattergram, or scatter plot. However, this graphical, as opposed to numerical, method can only be employed for showing correlations between two interval/ratio characteristics. Moreover, it can't be utilized to make trustworthy inferences.

It is very important not to confuse correlation with causation. Correlation is the degree to which states manifested by one characteristic coincide with states manifested by one or more other characteristics. But causation is the degree to which states manifested by one characteristic are due to—caused by—states manifested by one or more other characteristics. At least moderate correlation (a moderately strong relationship) must exist for causation to be claimed with any conviction to exist and thus for a relationship to be justifiably deemed causal. Yet moderate correlation (i.e., 0.41 through

0.60, roughly), even perfect correlation (i.e., 1.00), can exist without causation existing. Logically speaking, at least moderate correlation is a necessary condition, but not the sufficient condition, for causation and hence for a causal relationship to exist. Causation presupposes at least moderate correlation but even perfect correlation does not presuppose causation. Simply put, correlation is not causation!

8. Regression Equations: A regression equation is a mathematical description of and thus an equation describing a relationship between quantities exhibited by an interval/ratio characteristic (e.g., height) and quantities or categories displayed by one or more other characteristics (e.g., weight) of a group (e.g., American adults), at least one of these other characteristics being interval/ratio. Creating a regression equation is worthwhile only when the relationship it is to describe is at least strong, as evidenced by one or more high correlations (strong relationships) between the two or more characteristics constituting it.

3

Univariate, Bivariate, and Multivariate Analysis

Processing data into information is in essence analyzing data. Thus data processing and data analysis are essentially the same. Analysis here means separating, dissecting, or teasing apart data followed by rearranging data. Data analysis is done with mathematical techniques. If these techniques are statistical, data analysis is called statistical analysis. Statistical analysis can take place at three levels of complexity: univariate, bivariate, and multivariate. Univariate statistical analysis is the analysis of data on one characteristic of a thing, sample, or population. Bivariate (statistical) analysis processes data from two of their characteristics. Multivariate analysis analyzes data on three or more characteristics. The word variate means the data analyzed from a characteristic are treated as a variable. Variable is a label researchers apply to a characteristic whose variation, changes in state, or changes in how it manifests itself is what they want to learn about through research. Therefore, variable is a research term. Outside the context of research a characteristic is usually called a characteristic and not a variable. We'll use the terms characteristic and variable interchangeably in what follows.

From 1850 onward statistics became a tool of the big three: government (statistics bureaus), economics, and business. All three were interested in data on the population of a nation. Initially the data they analyzed were population data and not sample data. By 1900 more sample data than population data were being analyzed, thanks largely to the successful promotion of random sampling by statistically inclined economists during the late 1800s. Most analyses before 1900 were marginally statistical because statistics had little yet to offer in the way of useful techniques. Also most were univariate. Definitely more sophisticated concepts, along with the mathematical techniques for gearing them to data, commenced appearing during the 1890s, mainly in England.

This new wave in concepts and techniques improved greatly the caliber of univariate analyses and made bivariate and multivariate analyses more than pipe dreams. However, they were more potential than actual tools until approximately 1915. For it was not until then that they started to be more widely known and better understood by economists and businessmen as well as by sociologists, biologists, educators, psychologists, etc. Therefore, despite the fact that inferences about populations from samples started being made in the late 1800s, it wasn't until around 1915 that such inferences would begin to usually be founded firmly upon statistical concepts and techniques and thus qualify as bona fide statistical inferences.

Economists during the late 1880s commenced more accurately describing how states manifested by economic characteristics (e.g., prices) of things (e.g., commodities) vary with the passing of time or, put another way, vary as a function of time. They did this by constructing price indexes and learning which of the measures of average (e.g., means vs. weighted means vs. medians vs. geometric means) are best at revealing trends. They also did it by graphing 'moving' averages—ordinarily moving means—of time series. A time series is the occurrence of a series or string of things over some span of time. A moving average smoothes or reduces the jaggedness characterizing a curve in a time series graph, jaggedness tending to cover up trends which might be present in the curve. About 1900, economists began to make univariate statistical inferences that were parameter estimates and significance determinations.

A popular univariate inference was estimating the mean of a population. To be fairly accurate and to implement his level of confidence, the individual estimating a population mean had to calculate two things from his sample data. One was the sample mean. The other was the standard deviation of a probability distribution referred to as the sampling distribution of the mean. Now the standard deviation of the sampling distribution of the mean or, equivalently, the standard deviation of the mean is known as the standard error of the mean. It was conceived and first applied, to observations in astronomy, by Pierre Laplace, whom we met on page 11.

We need to clarify a few things before going further. First, there are many standard errors. Each of them is named after and specified in terms of a statistic: a sample mean, sample proportion, sample correlation (measured a particular way, such as measuring it via the Pearson product-moment correlation coefficient). Second, each is the dispersion in a statistic, measured via the standard deviation of that statistic. Hence a standard error is the standard deviation of a statistic. Third, a probability distribution created expressly to portray the dispersion—deviations—in a statistic is ordinarily more precisely termed a sampling distribution, the sampling distribution for that statistic. The word 'sampling' in sampling distribution indicates the distribution was created by drawing from a population of imaginary things countless samples containing a specific number of these things and therefore samples of a certain size. Sampling distributions exist for statistics as well as for things called test statistics. A test statistic (e.g., chi-square, t, F) is a statistic that's developed for the purpose of enabling us to determine—test—the significance of statistics (e.g., sample frequencies, means, variances).

Fourth, probability distributions, notably the normal and binomial distributions, often double as sampling distributions for statistics (e.g., for proportions, means, correlations). Fifth, the "error" in standard error reminds us of what statistics, particularly when making inferences, thinks of dispersion (page 11). Sixth, besides being employed for estimating parameters, standard errors commenced around 1900 to also be employed for determining significances—determining the significance of differences—involving one sample mean or proportion and a performance standard (e.g., less than 0.02, or two percent, of cattle arriving at Chicago stockyards are rejected). Such significance determinations exemplify univariate analysis.

The new wave in statistical concepts and techniques quickly led to closer ties between statistics and probability theory. Why? The reason is that by 1895 individuals astute in statistics knew statistical inference, chiefly determining the significance of differences, could not progress much further if confined to the probability distributions then existing, which were the normal and binomial distributions. Additional distributions would be necessary in order for statistical inference to accommodate an ever growing diversity in data and in questions being asked regarding data, distributions to which new inferential techniques could refer their calculations.

Now the normal distribution is applicable only to interval/ratio characteristics (page 27) from which thirty, preferably forty, or more pieces of data have been collected. And the binomial distribution is applicable only to characteristics that manifest no more than two states, say male vs. female or loud vs. not loud. Hence many data had no distribution upon which to base inferences. The situation as of 1900 would soon be worsened by a growing stream of data on ordinal characteristics pouring out of psychological research. Analyzing data from these characteristics was problematic. If the data were numbers, referring statistics calculated with them to the normal distribution was thought by some to be of dubious validity. And if the data were words specifying more than two states (e.g., low vs. medium vs. high motivation) then any statistic calculated employing them would be incompatible with both distributions.

The mystique which came to surround the normal distribution about 1815 commenced around 1885 crumbling under the weight of mounting evidence that despite its accurately describing dispersion in data from up to maybe 75 percent of interval/ratio characteristics it could no longer be seen as Gauss and Quetelet had seen it, as accurately describing the dispersion in essentially all interval/ratio characteristics. From 1885 onward the normal distribution was less often termed the error law, Gaussian law, or normal law, since law implies essentially 100 percent, not 75 percent, compliance. Old habits die hard, though, and not until roughly 1950 did textbooks generally replace these names with the name normal distribution. But economists, sociologists, and others in the 1890s familiar with statistics already knew well the normal distribution's shortcomings and looked forward to and even helped create the new probability distributions that would supply the inferential foundations required by new techniques for determining significance.

About 1895 Karl Pearson identified a dozen types of probability distributions to which data from interval/ratio characteristics could be inferred. Unfortunately these so-called Pearsonian frequency curves, even the simplest (the Type 3), were mind boggling to all except the mathematically adept user of statistics. Since such users were in the minority, the curves weren't used much. They were seldom referred to after 1945. A byproduct of Pearson's work on these probability distributions was a technique he devised in 1900 for assessing how well a frequency distribution of data conforms to them. This technique evolved into two of the new techniques for determining significance, the chi-square test of goodness of fit (conformity) and the chi-square test of independence or association. Chi is pronounced *ki*, as in kite. And chi-square is the name Pearson gave to the test statistic generated by his two techniques and to the sampling distribution for this statistic. These techniques are still utilized, to determine the significance of sample frequencies of categories of one nominal characteristic (goodness of fit) or two nominal characteristics (independence). Sometimes the techniques are respectively applied to one or two ordinal characteristics whose quantities (numbers) are treated like categories (words). The chi-square goodness of fit test is a univariate technique whereas the chi-square independence test is a bivariate technique.

Pearson's interest and thoroughness in describing frequency/probability distributions, correlation, and regression set the stage for what would from 1910 to 1950 be fashionable in statistical analysis. Skimming the statistics textbooks which appeared between 1915 and 1955 will reveal that goodly space was commonly devoted to indoctrinating readers in the new fashion: in measures of average, of dispersion, of a frequency distribution's lopsidedness (skewness), of a frequency distribution's peakedness (kurtosis), of simple correlation (correlation of two characteristics), of partial correlation (simple correlation with influence of a third, fourth, or perhaps fifth characteristic controlled), of multiple correlation (correlation of three or more characteristics), of simple regression (regression equations that relate two characteristics), of multiple regression (regression equations that relate three or more characteristics). Fashions change, however, and the foundation for what would next be in vogue was laid down from 1920 to 1940.

Its founders continued the English tradition (page 14). Chief among them were Ronald Fisher, Jerzy Neyman, and Egon Pearson, the son of Karl Pearson. But, around 1930, England had to begin to share the statistics limelight with America. For that's about when George Snedecor and Louis Thurstone got into the act.

The chi-square goodness of fit and independence tests of significance were the start of a long line of univariate, bivariate, and eventually multivariate inferential techniques for determining the significance of differences. Such techniques received shorter titles in the mid 1900s. During the 1940s they began being called significance tests and in the 1960s they began being also called hypothesis tests. Significance test was the title given them by Fisher. Hypothesis test originated with Neyman and Egon Pearson, whose hypothesis testing approach was an extension of Fisher's significance testing approach. Now when in the early 1890s Pearson commenced making statistics mathematically respectable there was little in the way of what today are termed, with equal correctness, significance tests and hypothesis tests. The only significance test in common use was the "z-test" employed to determine the significance of one mean (i.e., the significance of the difference between one sample mean and a performance standard). It utilized the standard error of the mean and was probably created by Laplace. In the next thirty years the z-test evolved into a more general technique that ended up taking several specific forms, principally the z-tests for one mean, one proportion, two means, and two proportions. But back in 1895 it took only the first form.

The z-test in any form has a drawback. It can be applied only if the data in a situation are relatively numerous and consequently the sample(s) in that situation are relatively large. By relatively numerous and relatively large means that a z-test can be used only if at least thirty, preferably forty, observations (data) were made (collected) and consequently there are at least thirty things in each sample studied. Hence a minimum of sixty data and things are needed to utilize it for determining the significance of the difference between two means or proportions. Thirty, even sixty, pieces of data are not usually thought by statisticians to be particularly challenging. Of course that is today, a time characterized by routinely doing calculations electronically with personal computers, desk calculators, or handheld calculators. This was not the case before 1960. Calculations before then were routinely done electromechanically with desk calculators or with paper and pencil, aided by a slide rule or set of mathematical tables.

Parenthetically, "z" is the symbol for z-score, or standard score, and standard score is a standard deviation unit. Standard scores are employed to specify degrees of dispersion in the standard normal probability distribution, which is what we've been calling the normal distribution. Whenever you see "z" in statistics or probability theory, automatically think "normal probability distribution". The rationale for referring the "z" calculated in a z-test to the normal distribution is thus. If the size of the sample or samples which enter into determining the significance of a sample mean or proportion or of a difference between two sample means or proportions is/are at least thirty, then the sampling distribution of this mean or proportion or of this difference between means or proportions is essentially normal, the calculated "z" is essentially the same as the "z" corresponding to it within the normal distribution, and we can therefore be assured the risk level that actually exists is essentially the same as the risk level we desire.

Now if a z-test is to be applied repetitively there'll soon be incentive to look for an easier test. An alternate significance test was devised in 1908 by William Gossett, an English chemist working for Guinness brewery in Ireland. His work seemed to entail frequently drawing a sample in order to monitor and thereby control the quality of brew manufactured. Besides the significance test he devised, Gossett defined its test statistic and the sampling distribution to which the statistic is referred. All three were dubbed "t". Hence the names t-test, t-test statistic, and t distribution. Like the first form of the z-test, the t-test was for determining the significance of one mean. It evolved into determining the significance of the difference between two means. Unlike the z-test, it never evolved a capacity to handle proportions. Guinness permitted Gossett to publish his thoughts on statistics, but only under a pen name. So he published using the name Student. Thus the occasional labels Student's t-test, t-test statistic, and t distribution. It's unfortunate that few knew about the t-test until around 1950.

The t-test was the earliest inferential technique applicable to small samples, though it could be and often is applied to large samples too. In statistical analysis, a large sample contains thirty or more things, with emphasis on more, and a small sample contains less than thirty things. The t-test was a welcome relief, once it was generally known. Now there was a prevailing opinion until about 1950 that a sample must contain a bare bones minimum of fifty things in order for the person performing a significance determination to have an acceptable level of risk. Pearson chided Gossett for utilizing small samples. Some of this opinion was due to the level of risk adopted before around 1950 frequently being very low, one percent or less. And some was due to the lack yet of a consensus on how to know the size sample needed to achieve a prescribed level of risk. Most followed, and still do, the rule of thumb: the larger a sample, the lower the risk. The result was the widespread practice, before about 1950, of employing a z-test only if there are a hundred or so things in a sample. Reinforcing the above view on the desirability of large samples was the fact that large to very large samples are characteristically needed when utilizing the chi-square test of independence.*

It would be timely for us to talk in some detail about a point sometimes glossed over in discussions of statistics; namely, the meaning and implications of independence or, more precisely, statistical independence. Nearly all the applications of descriptive and inferential techniques assume that the observations made and hence the data to which they are applied are independent of each other. Independence means no observation and no thing observed, the source of an observation, influences other observations regardless of whether all observations made in a study are of things in the one sample studied or in the two, or more, samples studied. Observations displaying independence are known as independent observations; observations not exhibiting it are called dependent (or related) observations. As go observations, so go the samples yielding them. Samples giving rise to independent observations are termed independent samples while samples giving rise to dependent observations are referred to as dependent samples. If samples are random, we speak of independent random samples vs. dependent random samples.

Independent observations are accomplished by joint deployment of two protocols in a study. One protocol is to observe each thing in a study one time only. The second is to

* The chi-square test of independence and the various coefficients of correlation are alike in two ways. First, it and they process data on two or more characteristics of things and hence are bivariate or multivariate techniques (applications of the test to bivariate and multivariate situations are respectively called simple and complex chi-square analysis). Second, it and they use one sample. The chi-square independence test requires large to very large samples. A sample of more than 50 things is needed when the contingency table, a type of frequency table, summarizing the frequencies into which data collected were processed has four entries (the smallest number possible), more than 200 things when this table has six or eight entries, and many more than 200 when it contains more than eight.

not only select a random sample but a random sample whose constituent things each had the same probability of being picked to be in the sample. For if things chosen to be in a random sample were not equally probable candidates then the observations eventually made cannot with full justification be deemed independent. Recall our definition of random sample on page 21. There we said a random sample is one out of all possible samples of a certain size that had the same probability of becoming the sample obtained. Now if a sample is a random sample then it conforms to this definition. And if a sample is an independent random sample then its constituents each had the same probability of being picked to be in the sample.

It's just as important for readers of research findings as for the researchers responsible for them to be alert to circumstances which are likely to produce *unintended* dependent random samples and thus unintended dependent observations. Some applications of the chi-square independence test in the past failed to satisfy the test's assumption that the data it analyzes are independent observations. Frequently the assumption was violated by each thing in a sample being observed more than once and therefore processing two or more observations per thing, not the one observation per thing assumed. Observations which are independent are said to contain no bias and to be unbiased observations; observations which are dependent are said to contain bias and to be biased observations. Whereas an unbiased observation is the result of random influences, a biased observation is the result of a mixture of random influences and systematic influences.

Not always are dependent observations and samples unintended! Sometimes, though not usually, they are intended. With all forms of the z-test the samples and observations employed are always assumed independent. And with the two forms of the t-test they're usually assumed independent. Yet there are times in experimental type research when the samples and observations employed in a t-test are assumed dependent and, consequently, a deliberate effort is made to make them so. The reason for utilizing dependent samples and observations is to control—eliminate—unwanted systematic influences that are likely to be exerted on observations by one or more characteristics of the things being studied. Such characteristics are called extraneous, nuisance, or lurking variables. Dependence is achieved through experiments designed around either of two principles of experimental design, repeated observations and matched observations. They are defined here in the simple form conducive to t-tests and their nonparametric cousins (to be discussed). But they also occur in a more elaborate form appropriate to certain other significance tests. Repeated observations is making two observations of each thing in one sample. Matched observations is making one observation of each thing in two samples, after first pairing a thing in one sample with a thing in the other sample so that both things (e.g., a Democrat and a Republican) in every pair of things created are matched—the same—regarding the state of a specific extraneous variable (e.g., religious preference). Of the two principles, matched observations is the one most often used with t-tests modified to handle dependent observations. Note that matching is typically termed blocking in a technique called the analysis of variance (page 49).

During the 1930s univariate analysis took a backseat to bivariate analysis as interest swung from describing a characteristic in isolation to describing the strength and details of its relationship with another characteristic. There was interest in multivariate analysis too, but this was formidable in a day when calculations were done with paper and pencil, slide rule, and mathematical tables. Describing the strength of and hence specifying the correlation for a multivariate relationship having three or four variables required doing a slew of calculations before ever actually calculating a Pearson product-moment multiple correlation coefficient. These preliminary calculations included sums, squares, sums of squares, and sums of cross-products. Just as intimidating were all the preliminary calculations needed to calculate the Pearson product-moment partial correlation coefficient for specifying the correlation for a bivariate relationship after unwanted influences of one or more other characteristics (extraneous variables) are controlled. Only a few people took on such grueling tasks, once in awhile. Very few did it routinely.

The last two important developments in univariate analysis took place about 1930. One was Ronald Fisher's criteria for good statistics for estimating specific parameters. The other was Jerzy Neyman's invention of a kind of parameter estimate known as an interval estimate or, to be more precise, a confidence interval estimate. It is one of two kinds of parameter estimates, the other being a point estimate. Let's look at both kinds, starting with point estimates. A point estimate is one number, a statistic. It is the only kind of parameter estimate possible if all a person knows is a statistic (e.g., a sample mean) related to the parameter of interest (e.g., a population mean). The problem with point estimates is they do not tell the person making them what level of confidence he ought to feel when saying a statistic (sample information) is essentially the same as a parameter (population information). For a point estimate to be of value, it should be a statistic calculated from data collected from a sample representative of the population whose parameter is being estimated.

A confidence interval estimate is two numbers, the lower and upper limits defining an interval presumed to contain the parameter of interest. Such an estimate is possible if a person knows a statistic (e.g., a sample mean) and the data used in calculating it. With confidence interval estimates the person making them is able to say with such-and-such level of confidence (e.g., 95 percent) the parameter of interest lies somewhere inside the above interval. Usually the parameter estimated via a confidence interval is a population mean or proportion. However, it can be a population correlation, standard deviation, etc. A confidence interval estimate is stated in terms of the name of the parameter estimated and the confidence level for the estimate. To illustrate, consider a hypothetical survey of the eating habits of seventh graders in New Jersey. One item in the report on the findings of this survey might be a confidence interval estimate worded "the 95 percent confidence interval for the mean daily intake of calories by seventh graders in New Jersey is 2400 to 2800". The confidence limits delineating the confidence interval of a confidence interval estimate should be calculated from the data collected from a sample representative of the population whose parameter is being estimated. Incidentally, the statistic related to data used in calculating the limits of an interval is at the center of that interval. This statistic is a point estimate. In the survey it is a sample mean of 2,600 calories.

On page 37 we said economists began about 1900 to make parameter estimates and significance determinations. Their parameter estimates were point estimates, particularly of means. If an economist wanted to be fairly accurate he would calculate the standard error of the statistic he based his point estimate on. Then when mentioning the estimate (e.g., the average length of a strike or lockout is fifteen days) in a conversation he would also mention this standard error (e.g., three days). By doing so, he communicated what until around 1955 was commonly termed the reliability of a statistic and thereby of the estimate made from it. It may appear a point estimate accompanied by its standard error is indistinguishable from a confidence interval estimate. Yet there is a subtle distinction between them. A point estimate and its standard error are two items, the first concerning a parameter and the second concerning a statistic. A confidence interval estimate is one item, the interval concerning a parameter. In essence Neyman united point estimates and their standard errors into confidence interval estimates.

In the decade preceding the above developments in parameter estimation Fisher was pioneering a versatile technique for determining the significance of differences between means by analyzing the variances associated with them. Said technique is the analysis of variance, or ANOVA. It was devised for use in experiments and consequently situations where most things that could exert unwanted influences on the things observed are, it is hoped, controlled (via randomization primarily, matching secondarily). The technique is employed to discover and describe causal relationships between characteristics of things. Causal relationships may be bivariate or multivariate. Hence analysis of variance may be bivariate or multivariate. If bivariate, it is called simple ANOVA. If mulitvariate, it is termed complex ANOVA. As an aside, we are here running into terminology similar to that met in discussing correlation, regression, and chi-square. Thus simple signifies two characteristics and multiple/complex signifies three or more characteristics.

No experiment arises magically. Time goes into formulating its purpose, designing it, settling on its methods and equipment, figuring out its cost, etc. before ever getting to the stage of performing it. Yet designing an experiment was simply something done and not in and of itself a subject to be studied, until the appearance of ANOVA. During the 1920s Fisher identified several generic designs of experiments to which his ANOVA could be applied. More were identified by him and others afterwards. Now researchers desiring to utilize ANOVA must decide upon which of these (generic) experimental designs to adopt so as to thereby know how to process data *before* data are collected, since how ANOVA calculations are done varies from one experimental design to the next. To make a wise decision, a researcher needs to know the design options available for use in discovering or describing the relationship between the two or more characteristics to be studied (the study variables) in a situation and for controlling extraneous variables identified in that situation.

Many ANOVA experimental designs exist. The better known ones are the completely randomized design (one-way analysis of variance) and randomized block design (two-way analysis of variance) used in simple analysis of variance and four factorial designs used in complex analysis of variance: the completely randomized factorial, randomized block factorial, Latin square completely confounded factorial, and split-plot or block-treatment confounded factorial designs. The variables (characteristics of things) in a design are subdivided into study variables and extraneous variables. Study variables are subdivided into one or more independent variables and one dependent variable. Whereas an independent variable is a cause in the causal relationship to be discovered or described by an experiment, the dependent variable is the effect in that relationship. All variables except the dependent variable are control variables, meaning they're under the control of a researcher. Control variables are subdivided into independent variables and extraneous variables (variables outside the relationship studied which nonetheless can affect it unless their influences are eliminated). A researcher controls an independent variable by means of manipulation or selection, manipulating it to make it exist in certain states or selecting states in which it already exists. The researcher controls an extraneous variable by means of randomization or maybe matching.

A few comments on terminology are desirable. First, randomization is also known as random assignment since it means assigning on a random basis, at random, or randomly the things in a sample to two or more groups created from the sample. It was mentioned on page 19 in our Whitman chocolates illustration. Second, in experimental research an independent variable is sometimes termed a factor; a state of an independent variable is sometimes called an experimental condition, factor level, treatment level, or treatment; and a state of the dependent variable is sometimes termed a treatment effect. Third, in non-experimental research (e.g., survey research) an independent variable is sometimes called a predictor variable, being that it is the basis for making predictions; the dependent variable is sometimes termed the predicted variable; and a state of the dependent variable is sometimes called a prediction.

Before getting into how the significance of differences between means is determined in analysis of variance we ought to make explicit something implicit thus far. How does a significant difference between statistics (e.g., two sample means) lead to the conclusion there is a relationship between characteristics? Now in our discussion of correlations and regression equations on pages 32 and 34 we learned they pertain to relationships between characteristics. Yet nowhere in our discussion of differences on page 32 did we see how they pertain to such relationships. We did learn they pertain to certain types of statistical information (frequencies, proportions, sets of ranks, averages, and dispersions) describing states displayed by some characteristic of interest (call it A). Left unsaid though was that there is implicitly another characteristic of interest (call it B); that B exhibits, let us say, two states; that A's two, say, sample means are related to B's two states; and that we will conclude there is a relationship between these two characteristics if we find a significant difference between the two sample means.

We ought to also note the names given significance tests. Some significance tests are named after their test statistics (e.g., chi-square test, z-test, t-test, F-test). Other tests are named after their inventors (e.g., the Wilcoxon test, Friedman's test, McNemar's test, the Cochran test). Still others are named after what, specifically, they analyze (e.g., signs, runs, sums, variances).

Analysis of variance analyzes variances associated with means in order to determine the significance of these means. It goes about analyzing variances by first partitioning, subdividing, or splitting the variance of the dependent variable in an experiment into all its sources. Its ability to do this is crucial and why ANOVA experimental designs are the way they are. Then one or more ratios are created, one with simple and two or more with complex ANOVA. A ratio is of two variances, actually two sources of variance. Next the quotient of a ratio is calculated. It is a test statistic which went unnamed until 1934, when George Snedecor, a great admirer of Fisher, named it 'F' in his honor. Finally the significance of the F, or quotient, calculated is found by comparing it with the appropriate F in the F distribution. If the calculated F is significant and thus deemed due presumably to more than random variation, the difference between the two variances entering into it is regarded significant and consequently the three or more means associated with one of these two variances (that in the numerator of a ratio) are, *as a whole*, judged significantly different. The italics stress the fact that among the three or more individual differences constituting a significant overall difference between three or more means there are likely to be one or more of them which are not significant and hence do not result from systematic variation in states displayed by the independent variable under consideration.

Snedecor was a professor at Iowa State University who became interested in statistics in 1914 and commenced applying it to agricultural research data about 1924. His early statistical work centered on multiple correlation and regression and on designing punch card machines to quickly categorize data collected. Snedecor was akin to Fisher. Fisher was interested in statistics and applied it to agricultural research data from 1919 to 1933, when he oversaw research at Rothamsted (agricultural) Experimental Station. Snedecor invited Fisher to visit his university in 1931 to orient researchers and students there to the then barely known analysis of variance, after which he championed its use in American biology. Not until the mid 1950s, however, did the technique start to be generally known and utilized in America, where it became fairly popular in psychological and educational experimentation. Of course it wasn't for an experimenter averse to calculations. Simple and particularly complex ANOVA demand more number crunching than the chi-square tests, z-tests, or t-tests, though not as much as Pearson's multiple and partial correlation coefficients. As an aside, the agricultural origin of the analysis of variance may be seen in the name split-plot given one of the ANOVA experimental designs, plot referring to a piece of land.

The last thing we'll say concerning Fisher is he did much to clarify and popularize the concept of the statistical significance of differences. It will be recalled from page 20 that (statistically) *signi*ficant differences are noteworthy. Why? Because they *signify*, or mean, sample information—some statistic—and population information—some parameter—are not essentially the same or, put another way, are essentially different. And a statistic and a parameter are essentially different if the difference between them is due to systematic variation, not random variation. It seems Fisher's view on (statistical) significance was a clarification of the thoughts about significance which appeared in the writings of Francis Edgeworth during the 1880s.

When data on an interval/ratio characteristic are normally distributed their dispersion, measured as the standard deviation, is construed thusly. A calculated standard deviation, say five, is the deviation from the mean, say forty, of 68 percent of these data. Call this standard deviation one standard deviation. Doubling and tripling it, we get two standard deviations (ten, or two times five) and three standard deviations (fifteen, or three times five). Inasmuch as these data are distributed normally and therefore constitute a normal distribution we automatically know from the properties of any normal distribution that one, two, and three standard deviations in the data respectively embrace 68, 95, and 99.7 percent of the data. Let us turn from this distribution of data to the distribution of some statistic. We can speak of one, two, and three standard errors of this statistic and thus of one, two, or three standard deviations of its sampling distribution. Also we can speak of one, two, or three standard errors of said statistic as respectively encompassing 68, 95, and 99.7 percent of instances of it that could be calculated were we to draw from some population all possible random samples of a certain size. However, we can say this only when working with certain statistics, primarily sample means and proportions, and with certain sample sizes, those thirty or larger.

The above percentages are related to probabilities. Specifically, the percentages 68, 95, and 99.7 percent correspond to the probabilities 0.68, 0.95, and 0.997. A probability states something will occur while a percentage states something did occur. Probabilities look to the future while percentages look to the past. Significance can look forward or backward. To avoid confusion, though, statisticians view significance as future oriented and having a counterpart termed risk which is past oriented. Hence they see significance as a probability and risk as a percentage. Significance and risk exist in degrees. We talk about levels of significance and levels of risk. Any level of significance can be adopted. Once a level is adopted the criterion for deciding some statistic is vs. isn't significant (is vs. isn't due to systematic variation) is specified. If in the case of the normal distribution we adopt a significance level of 0.05, then our criterion is the numerical value of "z" (i.e., 1.96) that coincides with the result of subtracting half this level (i.e., 0.025) from 0.500, said result being 0.475. Despite our discussion being couched in terms of the normal distribution, the gist of it extends to the binomial, chi-square, t, F, etc. distributions as well. Fisher soon settled on using 0.05 as the significance level in his research. About 1950 it became the standard of significance in most research.*

Offhand it would appear significance testing covers all that's relevant to determining significance. What more could be needed or wanted? In 1933 Jerzy Neyman and Egon Pearson suggested an improvement in Fisher's significance testing approach that became known as the hypothesis testing approach. Their approach started to attract a following in the early 1950s. During the early 1960s it largely superseded his approach. The result is that inferential techniques for determining significance, which in the 1940s began to be called significance tests, commenced during the 1960s being also called hypothesis tests. How do the significance testing and hypothesis testing approaches differ?

Testing the significance of a difference (between either a statistic and a standard or between two or more statistics) assumes that one of four possible outcomes will occur:

---

* A significance level (e.g., 0.05) and related risk level (i.e., five percent) are adopted at the time a study is being designed and are what a researcher desires for his *maximum* probability and percentage of being wrong if the decision he eventually makes is that such-and-such is significant. The practice in research reports is to give the adopted significance level, termed the "alpha level" or "alpha" (symbolized "α"), as well as the attained significance level, which on printouts made by statistical software packages used with personal computers is termed the "p-value".

1. deciding the difference is significant and it is significant,
2. deciding the difference is significant and it is not significant,
3. deciding the difference is not significant and it is not significant, or
4. deciding the difference is not significant and it is significant.

Significance testing takes into account the first and second outcomes and its significance/risk level refers to the second outcome and hence of being wrong deciding the difference observed is significant.

Significance testing sees the decision to be made as deciding between two hypotheses about a difference between population parameters corresponding to a difference between sample statistics. One hypothesis is a research hypothesis; the other is a null hypothesis. The research hypothesis, also called the alternative hypothesis (since it is the alternative to the null hypothesis), states there is a difference between the population parameters of interest. Depending on the purpose of a study, the difference will be non-directional (i.e., all differences less than and greater than such-and-such sized difference) or directional (i.e., all differences less than or greater than such-and-such sized difference). A research hypothesis is a composite hypothesis because it includes more than one size of difference. The null hypothesis states there is not a difference between the population parameters of interest. It is a simple hypothesis because it includes one size of difference; namely, no or zero difference. Only simple and hence null hypotheses can be tested!

This means that deciding between a research hypothesis and a null hypothesis must be based on the results of testing the null hypothesis. If testing a *null hypothesis* results in it being *rejected*, the *research hypothesis* is *accepted*. If testing a *null hypothesis* results in it being *retained*, the *research hypothesis* is treated as *unsupported*. Now testing a null hypothesis is achieved by testing the significance of a difference with a significance test. If that difference is significant, the null hypothesis is rejected. If it isn't significant, the null hypothesis is retained. Notice that it is common practice to say a null hypothesis is retained, not accepted, and a research hypothesis is unsupported, not rejected.

Neyman and Pearson started where Fisher left off. They took into account all four outcomes possible in a situation and then proceeded to explicitly view the situation as a decision-making problem wherein the researcher decides between two hypotheses, the null hypothesis and the research hypothesis. By its acknowledging the third and fourth outcomes on page 56 their hypothesis testing approach is able to thereby acknowledge and control and, in a few situations, even measure a second type of mistake, or error, a researcher can make. The first type of error is addressed in significance testing. It is the second outcome: deciding a difference is significant and it is not significant, which leads to rejecting a true null hypothesis and thus accepting a false research hypothesis. The second type of error, unaddressed in significance testing, is the fourth outcome: deciding a difference is not significant and it is significant, which leads to retaining a false null hypothesis and thus regarding a true research hypothesis unsupported. It's customary to refer to these types of error as Type I error and Type II error.

The very same decision is made in hypothesis testing and significance testing. It is to accept the research hypothesis vs. regard it unsupported by rejecting vs. retaining the null hypothesis in light of deciding a difference is vs. isn't significant. Hypothesis testing adds to the first potential error in making this decision a second and asks researchers to control it as best they can. A seesaw relationship exists between Type I and Type II errors. If all in a situation stays constant except for the probabilities of Type I and Type II errors and if the Type I error probability (termed alpha, the significance level adopted) is decreased (e.g., from 0.10 to 0.01), we find the Type II error probability (termed beta) will increase (e.g., from 0.70 to 0.84). One way to get a low alpha without an unacceptably high beta is to increase the number of things in the sample(s) studied—to increase sample size—by about fifty percent (e.g., from forty to sixty things per sample). Another way is to adopt a research hypothesis whose difference is directional instead of non-directional (as long as this can be sensibly done), thus allowing the alpha we want (e.g., 0.05) and a lower beta by permitting us to legitimately use the criterion associated with an alpha twice that size (i.e., 0.10). Still another way is to choose that significance test among the two or more applicable to the situation considered which is rated the most powerful, meaning the best at keeping beta small.

This might be a good place to say the criterion we mentioned on pages 23, 24, 26, and 55 is commonly termed the critical value in determinations of significance. It is called this because it's that value, that number, out of all the many values embodying a sampling distribution (e.g., the chi-square, t, or F distribution) or probability distribution acting as a sampling distribution (e.g., the normal distribution) to which is referred a test statistic calculated from sample data, a value which is dictated by the significance level underlying the risk level adopted. It's the basis for deciding whether this test statistic is vs. isn't significant and thus whether we ought to reject vs. retain our null hypothesis and thereby accept our research hypothesis vs. deem it unsupported. An example of a critical value is the "z" of 1.96 in our illustration at the top of page 55.

Among the various mathematical tables a person may consult are those containing the critical values associated with sampling distributions and aptly termed tables of critical values. The critical values in such tables are sometimes called tabled values since they are within the body of a table. Now when the normal distribution acts as a sampling distribution the critical values sought—the "z" values—are listed along the left side and top of its table and not within the body, which contains the probabilities (e.g., 0.475) related to these "z" values (e.g., 1.96). Tables of critical values aren't used quite as much today because statistical software packages generate whatever critical values are appropriate to evaluate the test statistics they calculate from sample data fed them.

The decision that a difference is vs. isn't significant and therefore is vs. isn't caused by systematic variation follows from comparing the magnitudes of a test statistic (e.g., t) and a critical value (t). Test statistics and critical values can be positive numbers or negative numbers. Positive numbers are implicitly preceded by a plus sign; negative numbers are explicitly preceded by a minus sign. Whether their numbers are positive or negative rests upon the sampling distribution involved and upon the difference in a research hypothesis being non-directional vs. directional. Some sampling distributions, such as the chi-square and F, have positive numbers. All others, such as the z (normal) and t, have positive and negative numbers.

If the distribution employed has positive numbers, the research hypothesis must state a non-directional difference. The test of the null hypothesis using this distribution is called a one-tailed test or one-sided test: it tells us to reject a null hypothesis when a test statistic is greater than a critical value. But if the distribution has positive and negative numbers then the research hypothesis may state a non-directional or a directional difference. The test of the null hypothesis utilizing this distribution is called a two-tailed or two-sided test with a research hypothesis whose difference is non-directional and a one-tailed test with a research hypothesis whose difference is directional. A two-tailed test tells us to reject a null hypothesis when a positive test statistic is greater than a positive critical value and when a negative test statistic is less than a negative critical value. A one-tailed test tells us to reject a null hypothesis when, depending on the direction of the difference, either a positive test statistic is greater than a positive critical value or a negative test statistic is less than a negative critical value.

Let us now show in its entirety the chain of events and line of reasoning that go into determining the significance of a difference. A test statistic incorporating the difference whose significance is to be determined is compared to a critical value. The comparison leads to the decision this difference is vs. isn't significant. Said decision leads to the decision the null hypothesis isn't vs. is true and therefore to rejecting vs. retaining it. And that decision leads to the decision the research hypothesis is vs. isn't true and hence to accepting it vs. considering it unsupported.

We will finish this lengthy discussion on significance testing and hypothesis testing with two comments. First, a sample statistic (e.g., the mean) is not compatible with a sampling distribution (e.g., the t distribution) and therefore cannot have its significance determined by being referred to a sampling distribution. That's why a test statistic (e.g., the t-test statistic) is necessary. For test statistics are compatible with sampling distributions and this enables them to serve as a medium through which sample statistics can, so to speak, communicate with critical values. Yet test statistics (e.g., a calculated t) do not correspond perfectly to the critical values (e.g., a critical t) in the sampling distributions (e.g., the t distribution) to which they are referred. The reason is the difference in how they are calculated. While test statistics are calculated from sample data, critical values are calculated via the equations which describe sampling distributions. Nevertheless the similarity of test statistic and critical value is close enough, as long as all assumptions of a significance test are satisfied. Second, the inferential techniques utilized to determine significance can with equal validity be called significance tests and hypothesis tests. For the significance tested by a significance test is that of the statistic (e.g., a sample correlation) associated with the parameter (i.e., a population correlation) that is stated in the null hypothesis tested by an hypothesis test.

The last thing covered in this chapter are the many statistical techniques often called nonparametric and occasionally called distribution-free. Nonparametric and distribution-free signify these techniques make fewer assumptions than others about parameters (e.g., population variance) or about the shape (e.g., normal) of the distribution of states of a population characteristic concerning which data are obtained. A researcher is less apt to run into problems with their assumptions being honored. Most nonparametric techniques were invented between 1935 and 1955 in response to a mounting need for techniques to process data on nominal and ordinal characteristics, and on interval/ratio characteristics when the assumptions made by existing techniques (e.g., the Pearson product-moment correlation, multiple correlation, and partial correlation coefficients; z-tests; t-tests; and analyses of variance) aren't upheld in a situation. Many were American innovations. A lot of the impetus behind developing such techniques was the stream of data on ordinal characteristics that began pouring out of psychological research around 1910.

Among the many nonparametric techniques are descriptive and especially inferential techniques. Most nonparametric inferential techniques are for determining significance and thus are significance/hypothesis tests. The first nonparametric techniques were the two inferential techniques Karl Pearson created in 1900, his chi-square goodness of fit and independence tests met on page 40, and a descriptive technique devised by Charles Spearman in 1904, his rank-order correlation coefficient. A thirty-year hiatus followed, whereupon a bunch were invented from the mid 1930s through mid 1950s. The better known ones are Fisher's exact probability test (1934), Friedman's test (1937), the runs test (1939), the Kolmogorov-Smirnov goodness of fit test (1941), Wilcoxon's test (1945), the sign test (1946), the Cramér Coefficient (1946), the Mann-Whitney test (1947), the median test (1950), the Kruskal-Wallis test (1952), and Goodman and Kruskal's gamma (1954). All but two of them, Cramér's coefficient and Goodman and Kruskal's gamma, are significance tests.

Once the label nonparametric was pinned to the above techniques, about 1950, there came the realization that the opposite of a nonparametric technique must be a parametric technique, as exemplified by Pearson's correlation coefficients and the z-tests, the t-tests, and analysis of variance. A very readable book by an experimental psychologist, Sidney Siegel, published in 1956 and entitled *Nonparametric Statistics*, did much to encourage using nonparametric techniques in psychological research and showing how they differ from parametric techniques. Around 1960 two factions began taking shape, proponents of and opponents to nonparametric techniques (particularly those which are significance tests). Probably the staunchest defense of nonparametric techniques was launched in a more technical book by another experimental psychologist, James Bradley, published in 1968 and entitled *Distribution-Free Statistical Tests*. The commotion over parametric vs. nonparametric techniques subsided during the mid 1970s as each was seen to have its rightful place in data analysis.

Parametric techniques assume their data come from observing states manifested by interval/ratio characteristics. Recall from page 28 an interval characteristic has states which are quantities we can distinguish between by saying one is less than or greater than another by such-and-such. A ratio characteristic is an interval characteristic which has a complete absence of quantity, symbolized by the number zero. Data on interval/ratio characteristics are typically measurements. Nonparametric techniques assume their data come from observing the states of either nominal characteristics or ordinal/interval/ratio characteristics. A nominal characteristic has states which are categories. An ordinal characteristic has states which are quantities we can distinguish between only by saying one is less than or greater than another.

Also parametric techniques and those nonparametric techniques applied to ordinal/interval/ratio characteristics assume their data are quantities exhibited by continuous, as opposed to discrete, characteristics and therefore can be, in theory, infinitely subdivided, despite our inability in practice to discriminate between very fine gradations in quantity. An instance of a continuous characteristic is weight. We can sensibly speak of 50, 50.6, 50.64, 50.644, 50.6447, etc. pounds even though the resolving power of scales does not usually exceed two decimal places (go more than two places to the right of the decimal point). The reason for saying the above techniques assume characteristics are continuous is they compare their test statistics to critical values in continuous sampling distributions.

Probability theory classifies probability distributions into continuous distributions and discrete, or discontinuous, distributions. The normal probability distribution and the chi-square, t, and F sampling distributions are continuous while the binomial distribution is discrete. The continuousness of the quantities processed by a technique is what is meant in statistics by the term continuity. Sometimes we hear of correcting for continuity or of a correction for continuity. This refers to making categories, which by definition are discrete, and discrete quantities, which are quantities treated as categories, more compatible with techniques (e.g., the chi-square tests) that refer their calculations to continuous distributions. Related to correcting for continuity is correcting for ties or a correction for ties. This refers to making any identical quantities spotted among data collected, which is likely with quantities that are ranks, more compatible with these techniques. Now if any technique, be it parametric or nonparametric, assumes characteristics are continuous and they are not then its descriptions or inferences will be less accurate.

A feel for many of the better known nonparametric techniques can be gleaned from looking at the kinds of characteristics and types of statistical information they address. Three techniques deal with nominal characteristics. The chi-square goodness of fit and independence tests are significance tests whose chi-square test statistics are calculated utilizing frequencies generated from data. Cramér's statistic is a measure of correlation which is calculated making use of the test statistic calculated for the chi-square test of independence. Various nonparametric techniques deal with ordinal characteristics and may be applicable to data on interval/ratio characteristics when the assumptions made by parametric techniques are not upheld. Goodman and Kruskal's gamma is a measure of correlation calculated with frequencies compiled from data; the sign test is a significance test which also relies on frequencies. Most other nonparametric techniques dealing with ordinal characteristics base their calculations on ranks assigned data (see "6" on page 32 for more on ranks). One is the Spearman rank-order correlation coefficient. The rest are significance tests: Wilcoxon's test, the Mann-Whitney test, the Kruskal-Wallis test, and Friedman's test. Some nonparametric significance tests, when performed in connection with large or fairly large samples, calculate "z-test statistics" which are then compared to "critical values of z" in the (table for the) normal distribution.

Until around 1960 it was generally agreed two assumptions made by two parametric significance tests that started to become popular during the early 1950s must be satisfied. The two assumptions are normality and homogeneity of variance. Normality asserts that data for calculating the statistic(s) used by the test in question are quantities exhibited by an interval/ratio characteristic that's distributed in a normal fashion in the population(s) of things bearing this characteristic. Homogeneity of variance asserts that the population variances associated with the population characteristic on which data are collected are the same or, put another way, that the two or more pertinent population variances are equal. The two parametric significance tests are the t-test and analysis of variance (its F-test). Bias due to lack of normality is essentially prevented by employing samples or groups containing at least fifteen things; bias resulting from absence of variance homogeneity is essentially avoided by using samples/groups that are the same size.

Around 1960, a more relaxed attitude sprouted toward meeting these two assumptions. By 1965 there was a feeling, though not a consensus, that the normality and homogeneity of variance assumptions of the t-test and the analysis of variance are largely unimportant. Both tests were touted as robust. Robustness meant a strong tendency for no difference to arise between adopted vs. actual probabilities of committing Type I errors and Type II errors when a specific condition (e.g., normality) that a specific parametric significance test (e.g., the t-test) assumes must be upheld is not upheld in the situations to which it is applied. If no difference occurs between the adopted, often known as the nominal, and actual error probabilities when a test assumption is violated, the test is said to be robust insofar as that assumption.

During the early and mid 1960s James Bradley did extensive research that confirmed the original belief in the importance of the above two assumptions. An excerpt from his book will show readers that statistics is not devoid of passionate moments. In speaking of the myth of robustness Bradley (1968, p 25) says:

> It is presumably attributable to a none-too-careful reading of robustness studies by statistical practitioners and to an incautious summarization of results by some of their authors and by the authors of statistical textbooks. It often takes the form of a statement that "the ____ test is robust against the ____ assumption," or, worse, that "the ____ test is robust." Such a statement represents pure semantic chaos. There is no commonly accepted definition of what constitutes robustness, no agreed-upon criterion which distinguishes between a condition of robustness and one of nonrobustness.—Nor does the statement specify the extent of the violation against which the test is alleged to be robust. But the "amount of robustness" tends to depend (among many other things—see below) upon the "amount of violation". Thus the only relevant variable mentioned in the statement is left unquantified.
>
> The most insidious thing about the Myth of Robustness, however, is that the "degree" of a test's robustness against violation of a given assumption is strongly dependent upon factors which are not involved in the statement of a test's assumptions, which are often not required in a complete description of the assumption's violation, and which are not mentioned in the usual allegation of robustness, as quoted above. These factors cause no distortion of Type I or Type II errors when all assumptions are met, but greatly influence the distortion occurring under a given violation of assumptions—i.e., the factors interact with whatever violation occurs. Examples of such factors are the following:

Bradley is advocating here for employing nonparametric significance tests in situations where before around 1960 parametric significance tests wouldn't have been used due to their normality or homogeneity of variance assumptions not being met. In truth, neither assumption *usually* needs to be satisfied perfectly. With few exceptions, the major one being when a directional research hypothesis is adopted, approximately normal distributions and roughly equal variances are sufficient, approximately and roughly meaning at least 75 percent.

4
Selective Elaborations

A lot of territory has been covered in the last three chapters. If you've understood the preceding material at least fairly well, you already know more than most people do about statistics. You now know more than just a little about the history of statistical thinking, what statistics tries to accomplish, and statistics as a branch of mathematics. To this may be added being more knowledgeable of samples/sampling methods, the commonly used types of statistical information, ways statistics has been applied, and the fundamentals of statistical inference, thereby knowing why statistics is closely tied to probability theory. No longer are you a stranger to such jargon as error, frequency distribution, correlation, statistic, parameter, confidence interval, significance test, critical value, risk level, null hypothesis, Type II error, etc. Yet you need to learn more before being truly literate in statistics. In this chapter are elaborated certain things previously discussed and certain new things, things the reader should by now be better able to understand.

Revisiting Randomness

We have seen that statistics is very interested in what in everyday language goes by the name chance and in mathematics and science is known as randomness, be it random influences (page 20), random samples (page 21), the random sampling procedure (page 23), random variation (page 25), independent/unbiased observations (page 45), or the randomization/random assignment principle (page 50). Interest in randomness is interest in random variables. Random variable is a term common to probability theory and hence to inferential statistics. A random variable is a variable—a characteristic—whose states are manifested at random in the things making up the population sampled, the population from which a sample is created. But what is randomness?

One could say random is chance. Accordingly, the random sampling of things in a population or the random occurrence of events is their chance sampling or occurrence. But this says no more than random and chance are synonyms. It's a tautology, a circular definition. A true definition is the following: random (chance) is the effect on things in a sample or on events taking place of many unknown, independently acting causes. This indicates randomness is something caused, not uncaused. Now a thing possesses various characteristics each of which exists in one out of two or more states at any moment. And an event is the occurrence of a state of a characteristic. If things in a sample are selected at random or if events happen randomly, the characteristics of things in the sample or the characteristic of these events become random variables. A characteristic may be said to ride piggyback on the thing possessing it. Therefore, whatever happens to this thing also happens to this characteristic.

Inferential techniques assume samples, and groups created from samples, are random in order to thereby assure the data they process are from characteristics which are random variables. The reason they assume this is so they can legitimately refer their test statistics to the critical values in sampling distributions. For critical values are states of a random variable. Hence characteristics of things are transformed into random variables by means of random sampling and randomization to uphold an assumption, that of a variable being random, which inferential techniques must make because the assumption is also made by sampling distributions and, in fact, by all probability distributions!

Usually it is easy to accurately create two or more random groups from one sample, be the sample random or otherwise. A mathematical table, dubbed a random numbers table, may be helpful in doing this. However, random groups are typically created by drawing well-shuffled slips of paper from some box or can. Each slip bears a number identifying a thing in the sample, and the box is shaken vigorously between draws. Usually it is not easy to accurately create a random sample from a population. If the population is finite as opposed to infinite in size, its things can all, at least in theory, be identified and listed in a sampling frame (page 21) and then things to be in a sample can be picked from said frame on a random basis with the help of a random numbers table. This process is prone to glitches requiring expertise to overcome. If the population is infinite in size, its things cannot all be identified and put on a list to pick from and therefore sampling is much less structured and takes on somewhat of a "by gosh, by golly" character. When a researcher needs a random sample and isn't sure she will end up with one, she can design her study around observing the sequence in which things are picked to be in the sample, conduct a significance test called the runs test on "runs" observed in this sequence, and decide from the test results whether the sample is probably random and hence whether the sample will meet the random samples assumption of the significance test(s) she plans to use.

Shopping for Significance

A significance/hypothesis test should be chosen at the time a study is being designed. That way the assumptions it makes can be better met through details (e.g., sample size) in the design of this study. Once data have been collected and screened (e.g., for normality and/or homogeneity of variance) it may be found they do not satisfy one or more of the test's assumptions. This is more likely if the test picked is parametric, such as the t-test, than if it is nonparametric, such as the Wilcoxon test. Yet if there is no evidence at the time a study is designed that data eventually collected on an interval/ratio characteristic are apt to run afoul of an assumption made by a parametric test then it is probably smart to plan on employing the parametric test because it will usually be at least slightly more powerful (page 58) than its nonparametric counterpart. Suppose after data are collected they are found to violate an assumption of the parametric significance test adopted. All that can then be done is select the best comparable nonparametric significance test and process the data using it.

What if the results of performing a significance test indicate the researcher ought to retain the null hypothesis and thereby treat the research hypothesis as unsupported? A research hypothesis is not something a researcher is indifferent to. Given the time, effort, and money going into a study, nonchalance on his part would be surprising and would no doubt alarm his employer or funding source. A research hypothesis is a pet hypothesis! It's something a researcher very much wants to end up accepting. Such a desire pressures a researcher into thinking of ways around having to conclude that his research hypothesis was not borne out by his data. These pressures may lead a researcher astray.

Three ways may be entertained for salvaging an unsupported research hypothesis. All are questionable ethically. One is to say in the report of a study that the significance level (originally) adopted was higher (e.g., 0.10) than it really was (e.g., 0.05), thus permitting the research hypothesis to not only be accepted but accepted at a risk level which readers of the report will wrongly think is the one originally believed desirable. However, people in the know who see an alpha level higher than what is customary for the sort of research being reported will suspect hanky-panky. Another way is to replace the non-directional research hypothesis originally formulated with a directional one, followed by writing the report as if this was the research hypothesis originally set forth. Recall from page 58 that a directional research hypothesis stands a better chance at being accepted than does a non-directional research hypothesis. Still another way is to reprocess the data with a second test, hoping it will reveal significance not revealed by the first test. Now if the results of the second test were to ever overturn the results of the first then a researcher would face a dilemma in what to report.

Measures of Average and Dispersion

A guiding principle in choosing a measure of average or dispersion, indeed a measure of anything, is to pick one that makes use of all the data collected on some characteristic of interest and reflects the nature—nominal, ordinal, interval/ratio—of this characteristic. We start with measures of average. Generally speaking, it is best to use the (arithmetic) mean as the measure of average for data on interval/ratio characteristics, to employ the median as the measure for data on ordinal characteristics, and to utilize the mode for data on nominal/categorical characteristics. The mean is the sum of all quantities in a group, divided by the number of quantities in the group; the median is that quantity in a group's arrayed quantities below which fifty percent of the group's quantities fall; and the mode is the category in a group having the largest frequency.

There are important exceptions to the foregoing generalization on utilizing the mean. First, the median should be used for data on interval/ratio characteristics distributed in a skewed, or lopsided, way. Second, in settings where it's sensible to find the mean of two or more means this mean ought to be a weighted mean when these means were calculated from different-sized samples or groups. If, as often happens, a person coming across two or more means doesn't know the sample/group sizes which entered into their calculation, it is unwise to add up, say, three such means and divide their sum by three to thereby get an overall, or grand, mean. Third, no measure of average is usually very meaningful and thus useful when data are from a distribution with more than one hump (i.e., a "bimodal" or a "multimodal" distribution) or no hump (e.g., a U-shaped distribution or a rectangular, also referred to as a uniform, distribution).

The mid-range (the sum of the smallest and largest observed quantities, divided by two) is an alternative to the mean that is sometimes plotted on control charts employed for monitoring industrial processes. It is a very rough, yet quickly calculated, measure. The harmonic and geometric means are measures of average used with data from only ratio characteristics and from them only under certain conditions. Both assume data are quantities which are positive numbers greater than zero. The harmonic mean in addition assumes quantities are rates of a specific kind and constitute a stationary time series: the rates didn't systematically increase or decrease over the span of time they were observed. And the geometric mean additionally assumes quantities constitute a nonstationary time series: the quantities did systematically increase or decrease over the span of time they were observed. When data honor the assumptions made by them the harmonic mean and geometric mean ought to be used instead of the mean. Parenthetically, a nonstationary time series is commonly known as a growth function if quantities consistently increase with the passage of time and as a decay function if quantities consistently decrease with time.

As for measures of dispersion, the standard deviation and the variance are used for data from interval/ratio characteristics having approximately normal distributions. The standard deviation (the square root of the variance) is utilized as a descriptive measure, because it's expressed in the same units of measurement as the data collected, and as a component of inferential techniques, being employed to calculate standard errors. The variance (the square of the standard deviation) is utilized as a component of inferential techniques, particularly the analysis of variance. Neither the standard deviation nor the variance is of much value when applied to data from skewed distributions.

The quartile deviation, known also as the semi-interquartile range, is ordinarily used with data on ordinal characteristics. It is occasionally employed with skewed data from interval/ratio characteristics too. Another measure for such data is the average deviation calculated with the median rather than the mean (Hays, 1994, p 189). Perhaps the best measure for skewed interval/ratio data is Tchebycheff's, Chebyshev's, or Tchebysheff's theorem, applied to percentages instead of probabilities. The percentage of interest is calculated thusly: the person calculating it decides on the smallest number of standard deviations worth of dispersion that are of interest (typically two or more), refers to this number as "k", squares k, divides one by k-squared, subtracts the resulting quotient from one, and multiplies the resulting difference by one hundred. According to the theorem, this percentage is the minimum percentage of data falling k standard deviations to either side of the mean. Simply put, it is the smallest that dispersion can, and hence will, be. This is good to know if a report you read neither gives the standard deviation for the data it refers to nor gives the data so you can calculate it. Hays (1994, p 194) says that when data are approximately normally distributed the smallest dispersion possible is specified more accurately by multiplying the above percentage by 0.44.

A couple of measures of dispersion exist for data on nominal characteristics. They are seldom used though. One is the variation ratio (Blalock, 1979, p 76). It is one minus the quotient resulting from dividing the largest or modal frequency (the frequency associated with the mode or modal category) by all frequencies. The other is the index of qualitative variation (Mendenhall et al., 1974, p 121), whose calculation is more involved. Our last measure is the coefficient of variation, employed with ratio characteristics. Unlike the other measures of dispersion discussed, it is a relative, as opposed to absolute, measure. It allows us to compare two or more standard deviations by dividing each by its mean, thereby making them into proportions (usually into percentages).

Dichotomies

Characteristics are sometimes classified as dichotomous characteristics (dichotomies) vs. polytomous characteristics (polytomies) or, equivalently, as binomial characteristics vs. multinomial characteristics. The prefixes "di" and "bi" signify two states and the prefixes "poly" and "multi" mean many—three or more—states. Dichotomies are classified as natural vs. artificial. Natural dichotomies are nominal characteristics that naturally exist in one of two possible states. An instance is gender, whose two states are the categories male and female. Artificial dichotomies are ordinal or interval/ratio characteristics that exist in nature in one of three or more, usually many more, possible states and which for convenience are treated as existing in two states. Consider vehicle mileage.

It is a ratio characteristic displaying innumerable quantities. Frequently its mileages are condensed down to low vs. high, transforming it into an ordinal characteristic. Another example is hardness. It is an ordinal characteristic naturally existing in at least ten quantities, the ten-point scale of mineral hardness invented by German mineralogist Friedrich Mohs in 1812, yet is often construed as manifesting two quantities, hard and soft. Still another instance of an artificial dichotomy is employee adaptability, which is probably no more than an ordinal characteristic. It exists in seven or so quantities, say very poor vs. poor vs. fairly poor vs. fair vs. fairly good vs. good vs. very good, but often is dichotomized as unsatisfactory vs. satisfactory.

Care must be taken with dichotomies. Two cautions are extremely important. First, sometimes a generally accepted natural dichotomy may not in fact reflect nature but in truth mirror our ignorance of the nominal characteristic dichotomized. Of course, this is more a concern of research than of statistics. Statistics does not get off the hook though if it becomes a means for legitimizing and thereby furthering ignorance. It is good for us to question once in a while a longstanding natural dichotomy to be sure we still feel it is valid. Second, in spite of the frequent desirability of employing artificial dichotomies we should realize that doing so runs contrary to a tradition in statistics and research. This is the habit of collecting and processing data in ways which preserve the nature—nominal, ordinal, interval, or ratio—of a characteristic. As said on page 28, interval characteristics are mostly conveniences in dealing with data from certain ratio characteristics, such as temperature. Understating the nature of a ratio characteristic by treating it as an interval characteristic is common and is a commonly accepted exception to tradition. However, understating the nature of an ordinal or interval/ratio characteristic by treating it as an artificial dichotomy is a greater understatement and consequently ought to be done more cautiously. The first thing to do is determine what will be accomplished and what will be obscured by dichotomizing a particular quantitative characteristic. If after mulling over these potential issues it is decided to dichotomize, the next thing to do is determine how best to conceptualize and label the two quantities.

More about Correlation

We talked about correlation and correlation vs. causation on page 32 and about simple, partial, and multiple correlation on page 40. Five more things about correlation will be covered here. First, the numbers produced by measures of correlation—by correlation coefficients—are not to be viewed as probabilities or proportions (relative frequencies)! The correlations 0.32, 0.51, 0.67, 0.84, etc. generated utilizing, say, the Pearson product-moment correlation coefficient are neither probabilities nor proportions therefore, even though they may look like them and, like them, range from 0.00 through 1.00. Second, all measures of correlation produce positive numbers, or numbers implicitly preceded by a plus sign (e.g., 0.84). Most measures of simple correlation where both characteristics are quantitative also produce negative numbers, meaning numbers explicitly preceded by a minus sign (e.g., −0.84). Now if a coefficient can produce positive or negative numbers then its positive numbers denote direct relationships (i.e., relationships where increases in the quantities exhibited by one characteristic coincide with increases in the quantities displayed by the other characteristic) and its negative numbers indicate inverse relationships (i.e., relationships where increases in the quantities of one characteristic coincide with decreases in those of the other). Depending on whether no sign or a minus sign is in front of them, the correlations generated by such coefficients are termed positive correlations and negative correlations respectively.

Third, there are marked differences among correlation coefficients in how they crank out correlations and in the nature of the data they process, which is traceable to the nature of the characteristics they address. Consequently it is imprudent to think a correlation of 0.75 generated by four measures of correlation—Pearson's product-moment correlation coefficient, the point-biserial correlation coefficient, the correlation ratio (eta), and the Cramér coefficient—is the same strength of relationship for all four measures. But they are enough alike in their mechanics that the same number generated by each will at least be very roughly the same strength (i.e., somewhere between 0.70 and 0.80).* Fourth, the statistical significance of and not the size of a correlation is what counts! The reason is a sample correlation of 0.75 doesn't always stand a better chance than one of 0.50 of being found significant, though usually it does. For while 0.75 is a high correlation that implies a strong relationship, it refers to a sample relationship and not the population relationship hypothesized to be yielding the sample relationship. As an aside, notice that a significant sample correlation of 0.75 does not mean the population correlation is 0.75. However, a confidence interval (page 47) erected for this population correlation utilizing this sample correlation and having confidence limits from 0.72 to 0.78 will quickly show these two correlations to be within the same ballpark.

Fifth, a correlation may occasionally be characterized as spurious, spuriously high, or spuriously low. Spurious means

* Guilford (1936, p 341) indicates Spearman rank-order correlations are equal to or less than Pearson product-moment correlations, being less than them by 0.02 at most (when a Spearman correlation is from 0.36 to 0.75).

false. Hence these words are warnings a person utters to alert other people to his thinking the correlation so labeled is suspect and ought to not be taken at face value by them. Naturally this is one person's opinion and may in itself be false. Spurious has been misapplied. A clear misapplication is when someone asserts a correlation is spurious in response to hearing someone else claim it shows a relationship is causal. The first person's assertion brings to mind the quotation from Bradley on page 66, where he says of an assertion about robustness "Such a statement represents pure semantic chaos." We know from our earlier discussion that correlation is nether causation nor evidence for causation in a relationship. Also we can safely assume statisticians and scientists know this. Consequently, labeling a correlation spurious because it is wrongly construed to imply a causal relationship is senseless. For it is not the correlation but its interpretation that is spurious. What's more, the label can be misleading, inasmuch as it can be mistook to suggest there exist correlations that all by themselves can be justifiably interpreted as indicating the presence of causation in relationships.

It seems more informative to talk about three degrees of appropriateness in regard to using the concept correlation and using correlation coefficients than to dichotomize their usage as true or spurious. So we'll adopt three degrees of appropriateness: inappropriate, questionable, appropriate. We would then say using the concept or the coefficients *as the sole basis* for judging a relationship causal vs noncausal is inappropriate, while adding it is appropriate to use a coefficient to specify the strength of said relationship as long as its assumptions are met. We'd say, too, it is inappropriate to construe numbers produced by coefficients as proportions or probabilities. Also we would say comparing correlations generated by different coefficients is questionable. If a person does this, he ought to do it cautiously and tell people the coefficient generating each correlation compared. Finally, it's questionable to calculate a simple correlation when knowing or believing one or more specific characteristics lying outside the relationship whose strength is sought affect this relationship and hence will affect the size of any simple correlation calculated for it, making that correlation spuriously high (overstating it) or spuriously low (understating it). A partial correlation is needed in such situations! If no partial correlation coefficient has its assumptions honored in a situation where a simple correlation is dubious, my advice is to calculate no simple correlation rather than a dubious one.

Proportions and Percentages Linked to Correlations

Many correlation coefficients are so designed that they endow the correlations they generate with a very useful property. Any one of their correlations, say 0.80, can be squared to make a proportion (i.e., 0.80 x 0.80 = 0.64) and the proportion can then be multiplied by one hundred to make a percentage (i.e., 64 percent). This percentage is a measure of relatively how much dispersion in states manifested by one characteristic of a relationship is determined or explained by the other characteristic(s) of the relationship. Now if this percentage is subtracted from 100 percent then a second percentage (e.g., 36 percent) results. It is relatively how much the dispersion in one characteristic does not stem from the other characteristic(s) of the relationship and consequently may, yet also may not, stem from one or more characteristics lying outside the relationship. Recollect that a characteristic which is external to a relationship and influences the relationship is termed an extraneous, nuisance, or lurking variable.

The value of these two percentages is they convey a more meaningful impression of the strength of a relationship and *may* tell us by relatively how much the relationship is influenced by extraneous variables. Too small a percent for the first percentage, say less than fifty percent, and thus too large a percent for the second ought to trigger reflecting on these percents. If the correlation was simple and if reflection suggests the percents stemmed partly from identifiable extraneous variables not controlled, it might be wise to collect new data. The data would be on characteristics of the relationship and on certain characteristics known or suspected of being extraneous variables. After the data were collected they could be transformed into either a partial correlation (if the desire is that the bivariate relationship remain a bivariate relationship) or a multiple correlation (if the desire is for the bivariate relationship to become a multivariate relationship).

Correlation coefficients producing correlations from which such percentages can be validly calculated are the Pearson product-moment correlation coefficient, the Pearson product-moment multiple correlation coefficient, the Pearson product-moment partial correlation coefficient, the correlation ratio, the point-biserial correlation coefficient (Hays, 1994, p 336), and the Spearman rank-order correlation coefficient (Mendenhall et al., 1974, p 374). The first percentage and its proportion are called the coefficient of determination with the Pearson product-moment correlation coefficient and the coefficient of multiple determination with the Pearson product-moment multiple correlation coefficient. The second percentage and its proportion are known as the coefficient of nondetermination in both cases. As for the correlation ratio, or eta, the first percentage and its proportion are termed eta (its 'e' sounds like the 'a' in ate) squared.

Akin to these squared correlation measures is a measure calculated from the t-test statistic and the samples sizes of a t-test applied to two means whose data come from independent observations (page 44). It is the estimate of omega squared:

the proportion of dispersion in quantities of an interval/ratio characteristic determined or explained by dispersion in, or simply the difference between, states of a dichotomous characteristic (Hays, 1994, p 337). It is obtained the following way. First, square the t-test statistic. Second, subtract the squared t-test statistic from one. The difference is the numerator in the formula for calculating the estimate. Third, subtract one from the sum of the squared t-test statistic and two sample sizes. The difference is the denominator in said formula. Fourth, divide the number in the formula's numerator by the number in its denominator. The quotient is the estimate of omega squared. It can go from 0.00 (negative quotients are treated as zero) through 1.00 or, multiplied by one hundred, from 0 percent through 100 percent.

Contingency Tables

Research done in certain social sciences, especially in sociology and political science, is often oriented toward discovering and describing relationships between two, three, or four characteristics which usually are nominal but sometimes are ordinal. It is common for ordinal characteristics in these studies to be treated as being nominal characteristics. Hence their quantities are commonly conceived as categories and identified using words (e.g., young, middle-aged, old) rather than numbers and words (e.g., 17, 48, 79 years old). Data collected on categories of nominal and ordinal characteristics are converted to frequencies, or counts. Each frequency is the number of times some particular combination of categories (e.g., male, oriental, and old) manifested by the characteristics studied (i.e., gender, race, and age) was observed in the things (i.e., people) studied and therefore was observed to jointly occur and to hence be a joint event. This is why each such frequency is more specifically known as a joint frequency. The number of joint frequencies is equal to the number of categories studied. If a study is interested in three characteristics which respectively display two, four, and three categories, the number of categories investigated and, as a result, the number of (joint) frequencies compiled is twenty-four: two times four times three.

After their compilation the frequencies are placed inside—in the body of—a type of frequency table termed a contingency table. A contingency table is a set of intersecting (crisscrossing) horizontal and vertical lines which form rows and columns that serve to distinguish one characteristic from another as well as one category from another. Their intersections form boxes called cells. Frequencies are inserted in these cells, becoming known as cell frequencies. The result is a set of four or more cross-classifications, cross-tabulations, cross-tabs, cross-partitions, or cross-breaks. Cell frequencies in each row are added together and their sum is put in a row's right margin while cell frequencies in each column are added up and their sum is placed in a column's bottom margin. In this way are produced the row and column margin totals of a table. The number of characteristics and categories depicted by the rows in a table and the number depicted by its columns are used to designate the table. A table with one row characteristic having three categories and one column characteristic having five is termed a "three by five (contingency) table" and a table with one row characteristic having two categories, one column characteristic having four, and one column characteristic having three is a "two by four by three table". The first table has fifteen cells; the second has twenty-four. Often a contingency table depicting two characteristics is referred to as a bivariate frequency table.

Contingency tables are best known for their usage in conjunction with the chi-square test of independence. When so utilized, each cell in a table has a small cell in its top right corner. In it is the expected frequency for a cell as opposed to the frequency compiled for the cell, which in all chi-square tests is termed the observed frequency. The chi-square independence test operates by determining if differences between observed and expected cell frequencies are large enough to indicate a significant difference does vs. doesn't exist between them. From this is decided whether a relationship does vs. doesn't exist between the characteristics (in the population). Incidentally, the chi-square test statistic calculated in a chi-square test of independence is sometimes employed to calculate the measure of correlation called Cramér's coefficient, an improvement upon the contingency coefficient devised by Karl Pearson.

It should be pointed out that contingency tables are used with other significance tests too as well as with measures functioning somewhat the way measures of correlation do. These measures generate proportions specifying by how much errors made in predicting the states of one nominal characteristic in a bivariate relationship are reduced by knowing (the states of) the other nominal characteristic, a concept which coalesced about 1960 and goes by the name proportional reduction in errors. Two proportional reduction in error, or PRE, measures are the tau—pronounced *tow*, as in towel—and the lambda invented by Goodman and Kruskal. Blalock (1979; pp 307–311, p 315) feels that tau is superior to both lambda and the Cramér coefficient.

Graphs of Relationships

Relationships between characteristics of things can be described verbally (with words) and also can frequently be described mathematically, meaning graphically (with graphs) or analytically (with equations). Mathematical descriptions are more precise than verbal descriptions. Analytical descriptions are more precise than graphical descriptions. There are drawbacks to mathematical descriptions however. One is they are of value only to people possessing at least eighth grade mathematical skills. Unfortunately, these skills wither when not routinely exercised. Another is the number of characteristics that can be addressed in a situation is fewer with mathematical than with verbal descriptions. This is especially true of graphical descriptions since they are confined by the three dimensions of space: a graph is largely limited to depicting three characteristics, just as a contingency table is largely limited to depicting four. Still another drawback is graphs and equations are generally more effectively applied to interval/ratio characteristics than to nominal/ordinal characteristics, although since the mid 1970s there's been much headway including nominal/ordinal characteristics in equations. In the present section we look at graphical descriptions of relationships between characteristics and thus at graphs of relationships. In the next one we look at analytical descriptions of relationships between characteristics and therefore at equations of relationships.

Before discussing graphs of relationships we ought to point out they should not be confused with graphs of frequency distributions. Recollect from page 28 a frequency distribution portrays frequencies of states of a particular characteristic of a group as well as frequencies of events associated with a group, events being occurrences of categories of a nominal/categorical characteristic. A frequency distribution can take the form of a table of frequencies—a frequency table—or the form of a graph, where frequencies are depicted either as bars (i.e., a bar graph or a histogram) or as lines drawn between plotted points (i.e., a line graph that here is termed a frequency polygon, due to its shape). Note that bar graphs and histograms both have bars, customarily arranged vertically. The two differ in that with bar graphs, used with nominal characteristics, the bars are separated by spaces and with histograms, used with ordinal and interval/ratio characteristics, they are not, thereby indicating continuity between quantities. In popular accounts the graph of a frequency distribution of a nominal characteristic is often a pie chart (circle graph) whose "slices" roughly mirror in size the percentages of the categories exhibited.

Graphs of relationships between two interval/ratio characteristics are plots of points representing quantities of one characteristic, listed along the vertical axis of the graph, that were found to coincide with quantities of the other characteristic, listed along the graph's horizontal axis. If two characteristics are perfectly correlated, lines are drawn between adjacent points or, more precisely, are drawn connecting adjacent points. The lines in effect give rise to a single line regarded as the curve describing the relationship between these characteristics. Actually, the "curve" is a straight line or a curved line! Confusion here is avoided by the name given a relationship. A relationship described by a curve that is a straight line is called a linear relationship and a relationship described by a curve that is a curved line is called a nonlinear relationship or, sometimes, a curvilinear relationship. If two characteristics aren't perfectly correlated, a single line must be drawn in such a way that half the points are above it and half are below it. This line, or curve, is said to be fitted to the points and should be the line which best fits them and therefore is what's known as the line of best fit. However, the goodness of the fit of curve to points achieved through visual inspection (eyeballing) is satisfactory only when at least a high correlation—strong relationship—exists between the characteristics.

Traditionally the characteristic represented by the vertical axis of a graph is called the graph's dependent variable or predicted variable while that represented by its horizontal axis is termed its independent variable or predictor variable. As the terms predictor and predicted suggest, the curve in a graph not only describes a relationship but supplies the means for making a prediction of the quantity manifested by a dependent variable based on the quantity manifested by an independent variable. If a relationship is known or felt to be causal, the practice is to treat the characteristic which is the effect as the dependent variable and the characteristic that's the cause as the independent variable. And if time is one of the two characteristics related in a graph then the practice is to make it the graph's independent variable and to refer to the graph and its curve as a time series. With a time series predicting, as such, is irrelevant and replaced by forecasting. Forecasting involves using information on quantities exhibited by a dependent variable (e.g., average annual loss of American job income "caused by" American companies moving manufacturing plants to other countries) during the interval of time covered in a time series graph (e.g., 1985 through 2005) to forecast—to project—what quantity will occur at some particular time in the future (e.g., 2010). Short time intervals may show steady increases (growth) or decreases (decay) in quantities of the dependent variable. Long intervals may show that an increase gives way to a decrease which then gives way to an increase, showing a time series that is cyclical or oscillatory.

Graphs of relationships between two interval/ratio characteristics occasionally have a third characteristic that is a second independent variable, a characteristic which need not be interval/ratio. If this happens the graph will have more than one curve. Each curve in it shows the relationship between the dependent variable and first independent variable, at a specific value of the second independent variable. Comparing the curves shows how this relationship is affected by the second independent variable. Ordinarily no more than three or four states of a second independent variable are graphed. Therefore a graph will usually have no more than four curves.

Equations of Relationships

Equations describe curves. Curves describe relationships. Hence equations describe relationships too. Also, the characteristics making up a relationship are the dependent variable and the independent variable(s) on a graph as well as in an equation. While the dependent variable and independent variable(s) on a graph are denoted by the names of the characteristics of the relationship addressed, the dependent variable is symbolized Y and the independent variable(s) is symbolized X in the equation of the graph's curve. If an equation has more than one independent variable, each X is given a number subscript standing for a specific independent variable. Relationships between characteristics are either functional relationships or nonfunctional relationships. Functional relationships are the most sought after of relationships because they possess a very desirable feature, which is that each state of an independent variable coincides with only one state of the dependent variable.

Various kinds of functional relationships or, simply, functions are distinguished. Each kind exists as numerous specific functions all of which are alike in being associated with similarly looking curves and similarly looking equations yet are different insofar as the specific numerical values of their dependent variables, independent variables, slopes, etc. The various kinds of functional relationships are dichotomized into linear and nonlinear (curvilinear) functions and nonlinear functions are classified as quadratic, polynomial, rational, exponential, logarithmic, etc. functions.

Some relationships between characteristics are functions. Of those that are not, many aren't because a less than perfect correlation exists between their characteristics. If a less than very high correlation exists between characteristics of a relationship, the slope(s) in the equation describing it ought to be calculated in a manner that reduces to a minimum the dispersions in its characteristics that are implied by a less than very high correlation. Such a slope can be calculated via the method of least squares, but only if the dependent variable and independent variable associated with the slope are linearly related and thus related to one another by a curve which is a straight line. A slope calculated utilizing the least squares method is the basis for creating a straight line of best fit termed a regression line and for creating an equation termed a linear regression equation. Like any equation describing a relationship between characteristics, a (linear) regression equation has one dependent variable, one or more independent variables, and one or more slopes. By the way, a slope is the rate at which a dependent variable changes in response to changes in an independent variable, said rate being constant in linear relationships.

Regression equations are subdivided into simple regression equations, which are used for describing bivariate relationships, and multiple regression equations, used to describe multivariate relationships. It is assumed in creating a simple regression equation that the equation's dependent variable and independent variable are interval/ratio characteristics of a linear relationship. It's assumed in creating a multiple regression equation that the dependent variable is an interval/ratio characteristic, at least one independent variable is an interval/ratio characteristic, and the relationship(s) between the dependent variable and the independent variable(s) is linear. Meeting the assumption of linearly related variables is necessary in order for the method of least squares to succeed at reducing to a minimum the dispersions in a relationship's characteristics when it goes about calculating the slope (s) in the regression equation for that relationship.

If the plotting on regular graph paper of data collected for a dependent variable and an independent variable produces a pattern of points indicating the relationship between the variables is nonlinear, the linearity assumption can nevertheless still be frequently upheld by transforming the relationship into a linear relationship by transforming the data on its dependent variable and/or independent variable. Four data transformations exist for this (Johnson & Bhattacharyya, 1992, p 483): make Y values into natural logarithms, convert Y values and X values into common logarithms, transform Y values into reciprocals, and make X values into common logarithms. The first two call for the Y-intercept term in an equation being similarly transformed. Which, if any, data transformation will *linearize* a nonlinear relationship is a matter of transforming data followed by plotting transformed data on regular graph paper to see what pattern emerges. Data transformations can be a headache. Once a slope is calculated from transformed data, consistency dictates that all values of the independent variable(s) which are plugged into an equation and all values of the dependent variable generated employing the equation be similarly transformed. With multiple regression equations a special problem arises. If one slope is based upon values of the dependent variable that were transformed one way and another slope is based upon dependent variable values that were either not transformed or transformed another way, values of the dependent variable cannot be found using the equation.

A multiple regression equation may have one or more independent variables which stand for one or more nominal, ordinal, or dichotomous characteristics in a relationship. Each state of each such characteristic is termed a dummy variable, or indicator variable, and each dummy variable is seen as existing in either of two states, absent (indicated by "0") and present (indicated by "1"). Consider the dichotomy sex. If sex is an independent variable in a multiple regression equation, each of its states, male and female, is seen as two dummy variables. Thus male would be the dummy variables male absent and male present while female would be dummy variables female absent and female present. As with the interval/ratio independent variable(s) in a multiple regression equation, each dummy variable is associated with a slope. This is as far as multiple regression goes in dealing with characteristics in a relationship which are not interval/ratio. Yet there is a technique that does go further in dealing with them whose appeal has grown since the mid 1970s (Kerlinger & Lee, 2000, pp 808–811). Said technique, logistic regression, is like multiple regression in how it works. Its "logistic regression equations" contain one nominal/ordinal dependent variable, one or more nominal/ordinal independent variables, and/or one or more interval/ratio independent variables.

Multiple Comparisons

The expression multiple comparisons comes out of a research context, principally an experimental research context. It pertains to what, statistically, is done after uncovering a significant overall difference between three or more statistics. Normally these statistics are sample means and the overall difference between them was determined significant by performing an analysis of variance (page 52). Occasionally they are sample rank-sums that were found, as a whole, to differ significantly by conducting the Kruskal-Wallis test or maybe Friedman's test. After determining that the overall difference between three or more statistics is significant it is possible and advisable for a researcher to inquire into which of the three or more individual differences comprising this overall difference are vs. aren't significant, since one or more may not be significant. Doing this entails using a technique specially designed to determine more than one significant difference at a crack and thus compare multiple differences or, in other words, make multiple comparisons or contrasts. The word specially here means the techniques under consideration make their comparisons without allowing the level of risk the researcher adopted to increase as he proceeds from one comparison to the next. Such inflation in his adopted risk level would take place were the researcher to perform these comparisons in the commonsensical way of doing t-tests after an analysis of variance, Wilcoxon tests after a Kruskal-Wallis test, or sign tests after a Friedman test, his level of, say, five percent before doing the first of his comparisons rising to ten percent, perhaps more, by the last one.

A number of multiple comparison techniques have been developed for use with the analysis of variance: the Newman-Keuls method, Tukey's honestly significant difference (HSD) test, the Scheffé test, making orthogonal comparisons, the Dunnett test, etc. Most of them utilize a modified t-test or a modified t distribution or confine comparisons made with the standard t-test to what are termed orthogonal comparisons. Readers wishing to learn more are referred to Glass & Hopkins (1996, pp 444–469) and Roscoe (1975, pp 309–322). Also there is a multiple comparison technique used with the Kruskal-Wallis test (Siegel & Castellan, 1988, pp 206–216) and one used with Friedman's test (Siegel & Castellan, 1988, pp 174–184).

Bayesian Statistics

Bayesian statistics arose in the late 1950s, as did Bayesian decision-making. Both saw as important the then new view on probability called subjective, or personal, probability. The existing view on probability, dating to the birth of probability theory around 1655, became known as objective probability. An objective probability is a number from 0.00 through 1.00 that states the chances such-and-such event will occur (in accordance with the laws of chance, or randomness). It is more specifically termed a classical or *a priori* probability when based upon a logical analysis of events and an empirical or *a posteriori* probability when based upon compiling the frequencies of events observed. A subjective probability is a number from 0.00 through 1.00 that states a person's belief such-and-such event will occur (in accordance with assorted information of varying quality which enters into this belief). Subjective probability was conceived in 1926 by the Englishman Frank Ramsey, mulled over and refined during the next twenty-five years, dignified in 1954 by the American Leonard Savage regarding it a foundation stone of statistics, and promoted in 1959 by the American Robert Schlaifer in a book showing how subjective probability can be applied to solving a number of manufacturing and business problems. Subjective probabilities are manipulated, logically and mathematically, in the same ways objective probabilities are manipulated.

Increasingly from the late 1950s onward subjective probabilities were inserted in the relationship between probabilities called Bayes' theorem which was invented about 1740 by Thomas Bayes, a Scottish clergyman and mathematician. When wisely applied, the theorem is an important tool. Unfortunately it had by 1930 fallen into disrepute because of more than a few flagrant misuses of it in the early 1900s. But around 1955 safeguards in legitimately applying it were identified and adopted. Soon afterward Bayes' theorem commenced to be seen in a new light, by advocates of subjective probability. They saw it as a respectable way of integrating subjective probabilities with objective probabilities so as to more accurately state the belief/chances such-and-such event will occur. Basically, the theorem revises—upgrades—an existing probability by reversing what is known as a conditional probability and making an inference about this reversal, a type of inference called Bayesian inference. The importance attached to Bayesian inference by advocates of a subjective probability approach to decision-making and statistics led in the 1960s to the approach being increasingly called Bayesian decision-making and Bayesian statistics. Now while Bayesian decision-making has enjoyed a large following since 1970 Bayesian statistics has not. In statistics it has been primarily applied to making confidence interval estimates of population means and proportions (see Miller & Freund, 1985).

Statistical Decision Theory

Paralleling the development of subjective probability was the evolution of decision theory, nowadays usually termed statistical decision theory. Decision theory is called a normative theory because it sets forth norms which prescribe—tell—a person what to do or what alternative action to take in a situation. In this respect decision theory is unlike most theories. Most theories, scientific or unscientific, are explanatory theories. They explain (states displayed by) a characteristic or the existence of a relationship between characteristics. A theory prescribing how to make optimal decisions and hence to make decisions which are the best possible in light of what someone knows about a situation seems traceable to the thinking of the brilliant American-Hungarian mathematician John von Neumann. He sketched in 1928 and elaborated in 1944 a theory that prescribes how a person can make optimal decisions when competing with another person and thereby being in a conflict situation, meaning a situation where one person's gains or losses affect another's gains or losses. Such decision-making situations resemble in their competition and conflict games played by two people or two teams and the interactions between two companies out to win the same potential clientele or two nations after the same resources. Therefore his theory was dubbed game theory.

One rule von Neumann identified as useful in a competitive situation for deciding on what action to take or, as put by game theory, what strategy to adopt was the minimax decision rule, or decision criterion. It became a decision rule of decision theory too. The minimax rule gets its name from its instructing a decision maker to take that action or adopt that strategy which according to an analysis of a situation will always result in the decision maker getting the minimum of the maximum undesirable consequences possible in the situation. Decision theory was pieced together during the 1940s by the American-Austrian mathematician Abraham Wald to address situations not marked by competition/conflict. In the late 1950s the theory started to rapidly stress the benefits of a decision maker adopting one decision rule in particular; namely, the maximizing expected value criterion. Since expected value is a concept of probability theory and, by extension, of statistics, emphasizing it led to decision theory becoming generally known as statistical decision theory. A couple of incidentals are worth noting. First, applying (statistical) decision theory to situations is called decision analysis. Second, during the 1970s and 1980s a few textbooks on statistics, mainly those relating it to business or economics, interpreted the field of statistics in terms of decision-making.

5
On Using Statistics

In this, the final, chapter are offered insights and tips on using statistics. Learning them will put the finishing touches on making the reader truly statistically literate. We will be dealing here more with inferential statistics than descriptive statistics and more with significance determinations than parameter estimates. Of course this is in keeping with statistics being more interested in describing populations via samples than samples per se and more interested in describing population relationships between characteristics than population characteristics in isolation. In what follows is information I think is worth knowing by anybody planning on using statistics. By using statistics is meant applying one's knowledge of statistics to understanding the ever-increasing statistical information on economic cycles, medicine, product safety, crime, opinion polls, poverty, weather forecasts, advertisements, better education methods, etc. we learn about from television, newspapers, magazines, books, reports, journals, etc. If a reader desires to learn more about statistics, the next step is to study an introductory college textbook on statistics. I found one by Mario Triola (2006) particularly good.

Don't be timid asking for clarification of statistical information someone gives you on something. The silly question is the unasked question. Now statistical information may be vague because a person purveying it is so familiar with it he or she forgets others may not be. Or its purveyor may intend it to be vague in hopes of leading you to think along lines he wants you to think, in a direction that unbeknownst to you is either not supported or poorly supported by this information. Unscrupulous individuals realize that most people are simultaneously impressed with and intimidated by numbers, and so they relish conveying numerical information. Watch out for overly precise data and statistics, whose precision is unwarranted and deceptive.

Now it is important in doing the string of calculations entering into calculating some statistic, say a mean, to not round off to less than four decimal places—to less than four places after its decimal point—any component calculation except for the last component calculation, which results in and is the statistic calculated. The statistic itself, however, should not be rounded to more than one decimal place beyond the number of decimal places in the measurements utilized to calculate it, unless it's to be an ingredient of later calculations. Otherwise the statistic will appear more precise than it really is! Suppose you calculate the mean of the weights of ten people each of whom has been weighed to the nearest pound. It would be proper to give their mean weight as 185.6 pounds but not as 185.64, 185.646, or 185.6469 pounds. Proportions and correlations are exceptions in the rounding off of statistics. Both are always a zero followed by a two decimal place number (e.g., 0.72).

Sample size affects statistical inferences. It is a more important consideration with parameter estimates than with significance determinations however. For an increase in the size of a sample improves the quality of the parameter estimate obtained. In the case of interval estimates, a relatively large sample produces a relatively small interval within which the parameter being estimated is likely to occur. Formulae exist that indicate how large a sample is needed to construct a confidence interval no wider than such-and-such for a confidence level of such-and-such. Now the higher the confidence level desired the wider and thereby less meaningful will be the confidence interval created, unless sample size is increased noticeably to counter this widening. A sample of a few hundred things is commonly encountered with parameter estimates. With significance determinations, a sample is vastly smaller, ordinarily numbering fifteen to forty things. This is as it should be. For too large a sample can result in statistical significance (i.e., the difference is not caused solely by random influences) devoid of practical significance (i.e., the difference is not large enough to be of any interest).

Occasionally the population discussed in the report on a study is not the population from which the study got its sample(s). Either of two reasons typically account for this. One reason is mistaken identity: the population sampled was presumed to be A but was actually B. Misidentifying a population can happen fairly easily when the presumed and actual populations are similar, though not identical, in the things making them up, which can occur with sampling a human population if a sample is assembled by consulting an incomplete or outdated sampling frame (page 21). Misidentification also happens when the presumed and actual populations have a characteristic relevant to the study conducted that manifests states which vary systematically with the passage of time, this causing a population existing, say, ten years ago—the presumed population—to evolve into the actual population and thus cease to exist. Incidentally, any parameter estimated from a statistic on such a characteristic varies with time and is called a "time-varying parameter". The other of the above two reasons is the population of interest, known also as the target population, may be difficult to sample and consequently the researcher instead samples a similar one that's more accessible, with the idea of generalizing what is learned about it to the unsampled population. Experiments drawing convenience samples from college student populations typically generalize from a college student population to a broader human population, their target population.

If data are processed using the Spearman rank-order correlation coefficient, Wilcoxon test (also called his rank-sum test to distinguish it from his less often employed "matched-pairs signed-ranks test" or signed-ranks test), Mann-Whitney test, Kruskal-Wallis test, or Friedman test and if ties are found in the ranks created, it is usually desirable to apply an appropriate correction for ties, or ties correction factor, when performing the calculations. By a tie in ranks, or tied rank, is meant the same rank number, say five, is assigned more than once to the data collected. The reason for employing a correction for ties is that it offsets a loss in continuity caused by ties. More than one or two tied ranks will overstate a Spearman correlation, making it spuriously high. And more than one or two tied ranks will understate the test statistic of and thus reduce the chances of obtaining a significant difference with the Wilcoxon test/Mann-Whitney test (when ties exist *between* the two samples observed, as opposed to *within* them), Kruskal-Wallis test, and Friedman test. Some introductory statistics texts that discuss these techniques do not discuss their ties correction factors, which are well covered by Siegel and Castellan (1988).

On page 33 we said correlation is not to be confused with causation. Yet sometimes people don't think the edict applies to partial correlation. Their reasoning goes like this: partial correlation exercises control over (controls for) extraneous, nuisance, or lurking variables which might affect the strength of a relationship between two characteristics, removing any influence an extraneous variable might exert on one or the other of these characteristics. Control of extraneous variables is needed to demonstrate the existence of causation in a relationship. Therefore, partial correlation can demonstrate whether a bivariate relationship is causal or noncausal. But it can't! For it exerts no control over either characteristic embodying the bivariate relationship. Hence it is utterly unable to tell us which of the two is the cause of the other. The control needed to tell us this must take the form of physically manipulating that characteristic in a bivariate relationship we think is the cause in this relationship, such control being felt by most researchers to be attainable only through experimental research. Partial correlation has many legitimate applications. It's just that identifying causal relationships isn't one of them. We should point out that partial correlation and analysis of covariance are important techniques for exercising statistical control, a supplement to physical control.

Never lose sight of the fact the techniques of statistics have their limitations. First, they are limited by the situations in which they can be validly utilized. These limits are frequently stated as the assumptions they make about a situation. Inferential techniques make more assumptions—have more limitations—than descriptive techniques. This is especially true of inferential techniques devoted to determining significance, techniques generally known as significance tests and hypothesis tests. Nonetheless the assumptions of descriptive techniques or, equivalently, of statistics they generate, while rarely bearing the word assumption, are important too. The mean should not be utilized as the measure of average for data from an interval/ratio characteristic whose quantities are describable as a skewed (frequency) distribution. Seen from a different angle, the mean is limited to, or permissible with, an interval/ratio characteristic whose quantities can be described by a bell- or mound-shaped distribution. To this we can add that the Pearson product-moment correlation coefficient assumes data come from two interval/ratio characteristics and that a scatter diagram (page 33) shows these characteristics are at least close to being linearly related. If the diagram clearly shows a nonlinear relationship between them, the strength of their relationship should be gauged employing the correlation ratio.

Honoring assumptions can sometimes be somewhat of a juggling act. After checking for satisfaction of assumptions of a descriptive technique the next thing will typically be to check satisfaction of assumptions of an inferential technique, since interest is primarily in knowing the parameter associated with the statistic a descriptive technique produces. Meeting the assumptions of significance tests can be looked at this way. The two, three, or more assumptions made by a test cover all the ways, except one, that it will react to a situation by announcing a significant difference exists. By assuring all its assumptions are met we narrow the cause triggering the announcement down to just one; namely, the statistic (or the null hypothesis about the parameter corresponding to it) being tested. A second thing to be mindful of with statistical techniques is they are limited in how much their calculations can say. Consider regression equations. What an equation can say is limited by the observed quantities of the one or more independent variables which went into creating it. To generalize to quantities smaller than the smallest observed quantity and/or larger than the largest observed quantity is to say the relationship described by the equation holds—is the same—for these unobserved quantities as well, an assertion that's almost always unwarranted.

Perhaps the sole blemish on his otherwise sterling career was Karl Pearson's tendency to think flawed data can usually be rendered flawless through the judicious application of statistical techniques. This prompts a final remark on the limitations of these techniques, which is they are very limited in their ability to salvage flawed data and consequently are largely subject to the principle "garbage in, garbage out". Hence it is very important that the process whereby data will be collected is thought out carefully before actually starting to collect data! Pitfalls to avoid include failing to calibrate beforehand measuring devices needing periodic calibration, not assessing beforehand the reliability and validity of questionnaires or interview schedules, not training beforehand followed by supervising people tasked with collecting data, and not determining beforehand the expertise and reputability of individuals or organizations asked to supply data.

The last paragraph introduced the reader to considerations that involve both statistics and methodology. Methodology is applying various methods of varying complexity to research, whether scientific or unscientific. In its eyes, statistics is the chief method for processing research data. Herein lies the link between statistics and methodology. There are two methodological matters, or matters regarding method, to be aware of when doing a research study. Both can influence how its statistics are interpreted. One is the possible occurrence of interaction and the other is the possible occurrence of confounding.

Occasionally it's said that there is an interaction between two or more (independent) variables, or one (independent) variable interacts with another (independent) variable, or two or more (independent) variables interact with each other or are interacting. By this is meant that the influence which one independent variable exerts on a dependent variable is itself influenced by the influence exerted on the dependent variable by one or more other independent variables. An interaction between independent variables does not permit a researcher to predict the state of the dependent variable from knowing the state of each of these independent variables. In experimental research the effect on a dependent variable of an interaction between a combination of independent variables, termed an interaction effect, can frequently be isolated from the effect on it of an independent variable, termed a main effect, by the way an experiment is designed. Confounding is obscuring or hiding the influence an independent variable exerts on a dependent variable by letting changes in states of the independent variable coincide with changes in the states of another variable, called a confounding variable. The independent variable is said to be confounded with or by this other variable, or to be confounded. Although a confounding variable may be an independent variable, it is usually an extraneous variable that hasn't been controlled. In experimental research, confounding can be eliminated and normally is eliminated. There are a few experimental designs, however, that allow it to take place, under the conditions they stipulate.*

Methodological matters are tangential to our interests here. Individuals desiring or needing to know statistics will ordinarily have little desire or need to know a lot about methodology. Now if such knowledge is sought then it ought to be acquired in respect to the subject area concerning which it will be applied, say survey research, social research, behavioral research, educational research, or policy research (program evaluation). For, unlike statistics, books on methodology irrespective of subject area and hence not geared to a subject area

---

* The bimodal and multimodal frequency distributions alluded to on page 73 exemplify obscuring but not confounding. Despite their names implying these distributions have two modes or more than two modes respectively, the names are actually applied to any distribution with a mode (i.e., category having the largest frequency) and one or more categories, a few categories away from the mode, whose fairly large frequencies show up as one or more additional humps, or peaks, in the graph of the distribution. Bimodal and multimodal distributions are commonly interpreted as being frequencies compiled from and describing two groups or more than two groups, rather than, as should be the case, one group. What is obscured in them are the frequencies of the categories that the groups share, which are the categories falling in between the humps.

are essentially nonexistent. The reason is matters of method are, in their details, largely specific to the contexts within which research occurs. Were a book to be written on methodology in general which strove to mimic the approach taken by books written on statistics in general, it would probably run no more than approximately thirty worthwhile pages.

If this is a reader's second of the two thorough readings recommended in the Preface, he or she should now be literate enough in statistics to make sense of statistics and thus be better able to understand the statistical aspects of information encountered. As said at the beginning of the chapter, a particularly good place to go to learn more about statistics is Triola's text. One advantage a statistically literate person has over a person that isn't is an ability to communicate reasonably well with experts on statistics, with statisticians and with researchers well versed in statistics. This ability is indispensable to the person who must contact on her or his employer's behalf a statistician for advice or to be a consultant. Readers wishing to find statisticians specializing in certain kinds of statistical techniques (e.g., regression analysis, analysis of variance, Bayesian statistics, process control charts, time series analysis, statistics appropriate for stratified or cluster samples, and sequential analysis) might consider logging onto the American Statistical Association's website.

References

Blalock, Hubert M. Jr. *Social Statistics*. New York: McGraw-Hill, 1979 (revised second Edition).

Bowley, Arthur L. *Elements of Statistics*. New York: Charles Scribner's Sons, 1907 (third edition).

Bradley, James V. *Distribution-Free Statistical Tests*. Englewood Cliffs (N.J.): Prentice-Hall, 1968.

Campbell, John L. *Introduction to Science and the Scientific Method*. New York: Vantage Press, 2008.

Glass, Gene V. and Hopkins, Kenneth D. *Statistical Methods in Education and Psychology*. Boston: Allyn & Bacon, 1996 (third edition).

Guilford, J.P. *Psychometric Methods*. New York: McGraw-Hill Book Co., 1936 (first edition).

Hamburg, Morris *Statistical Analysis for Decision Making*. New York: Harcourt, Brace & World, 1970 (first edition).

Hays, William L. *Statistics*. New York: Holt, Rinehart & Winston, 1994 (fifth edition).

Johnson, Richard A. and Bhattacharyya, Gouri K. *Statistics: Principles and Methods*. New York: John Wiley & Sons, 1992 (second edition).

Kerlinger, Fred N. and Lee, Howard B. *Foundations of Behavioral Research*. New York: Harcourt College Publishers, 2000 (fourth edition).

Meitzen, August, *History, Theory, and Technique of Statistics*. Philadelphia: American Academy of Political And Social Science, 1891 (translation into English by Roland P. Falkner of original, in German, published in 1886). Interesting perspective of a statistician five years before statistics commenced becoming more mathematical.

Mendenhall, William; Ott, Lyman; and Larson, Richard F. *Statistics: A Tool for the Social Sciences*. North Scituate (Massachusetts): Duxbury Press, 1974 (first edition).

Miller, Irwin and Freund, John E. *Probability and Statistics for Engineers*. Englewood Cliffs (N.J.): Prentice-Hall, 1985 (third edition).

Porter, Theodore M. *The Rise Of Statistical Thinking 1820–1900*. Princeton (N.J.): Princeton University Press, 1986.

Roscoe, John T. *Fundamental Research Statistics for the Behavioral Sciences*. New York: Holt, Rinehart and Winston, 1975 (second edition).

Rudgley, Richard, *The Lost Civilizations of the Stone Age*. New York: The Free Press, 1999.

Secrist, Horace, *An Introduction to Statistical Methods*. New York: The Macmillan Co., 1917 (first edition).

Siegel, Sidney and Castellan, N. John Jr. *Nonparametric Statistics for the Behavioral Sciences*. New York: McGraw-Hill, 1988 (second edition).

Triola, Mario F. *Elementary Statistics*. Boston: Pearson Education (Addison-Wesley), 2006 (tenth edition).

Waugh, Albert E. *Elements of Statistical Method*. New York: McGraw-Hill Book Co., 1943 (second edition).